GCSE
Success

Maths

Higher Tier

Revision Guide

Deborah Dobson,
Phil Duxbury,
Mike Fawcett and
Aftab Ilahi

Contents

Adding and subtracting integers

Adding a **negative** number **or** subtracting a **positive** number will have the **same result**.

Adding a positive number **or** subtracting a negative number will have the **same result**.

$$3 + -5 = -2$$
$$3 - +5 = -2$$

Go down by 5.

$$-1 + +4 = +3$$
$$-1 - -4 = +3$$

Go up by 4.

Use a number line to visualise the answer.

Adding a negative number means subtract.

Subtracting a negative number means add.

+ +	means	+
+ −	means	−
− +	means	−
− −	means	+

Multiplying and dividing integers

Look at these examples.

Multiplying a negative number by a positive number always gives a negative answer.

$$-5 \times +3 = -15$$
$$+5 \times -3 = -15$$

Multiplying two positive numbers **or** multiplying two negative numbers always gives a positive answer.

$$+4 \times +3 = +12$$
$$-4 \times -3 = +12$$

The same rules work for division.

$$+10 \div -5 = -2$$
$$-10 \div -5 = +2$$

This table summarises the rules:

+	× or ÷	+	=	+
+	× or ÷	−	=	−
−	× or ÷	+	=	−
−	× or ÷	−	=	+

A positive number multiplied by a negative number gives a negative answer.

A negative number multiplied by a negative number gives a positive answer.

KEYWORDS

Integer ➤ An integer is a whole number; it can be positive, negative or zero.

Positive ➤ A number above zero.

Negative ➤ A number below zero.

Use of symbols

Look at the following symbols and their meanings.

Symbol	Meaning	Examples
$>$	Greater than	$5 > 3$ (5 is greater than 3)
$<$	Less than	$-4 < -1$ (-4 is less than -1)
$\geqslant$	Greater than or equal to	$x \geqslant 2$ (x can be 2 or higher)
$\leqslant$	Less than or equal to	$x \leqslant -3$ (x can be -3 or lower)
$=$	Equal to	$2 + {+3} = 2 - {-3}$
$\neq$	Not equal to	$4^2 \neq 4 \times 2$ (16 is not equal to 8)

Place value

Look at this example.

Given that $23 \times 47 = 1081$, work out 2.3×4.7

The answer to 2.3×4.7 must have the digits 1 0 8 1 ← *Do a quick estimate to find where the decimal point goes.*

2.3 is about 2 and 4.7 is about 5. Since $2 \times 5 = 10$, the answer must be about 10.

Therefore $2.3 \times 4.7 = 10.81$

Write the following symbols and numbers on separate pieces of paper.

| + | – | × | ÷ | = | 0 |

| +2 | –2 | +4 | –4 | +8 | –8 |

Arrange them to form a correct calculation.

How many different calculations can you make? For example:

| +2 | – | –2 | = | +4 |

1. Calculate the following:
    (a) $-5 - -8$
    (b) $-2 + -6$
    (c) $-7 + -3 - -5$
2. Calculate the following:
    (a) -12×-4
    (b) $24 \div -3$
    (c) $-3 \times -4 \times -5$
3. State whether these statements are true or false.
    (a) $6 < 3$
    (b) $-4 > -5$
    (c) $2 + -3 = 2 - +3$
4. Given that $43 \times 57 = 2451$, calculate the following:
    (a) 4.3×0.57
    (b) 430×570
    (c) $2451 \div 5.7$

Highest common factor

The highest common factor (HCF) is the highest **factor** shared by two or more numbers.

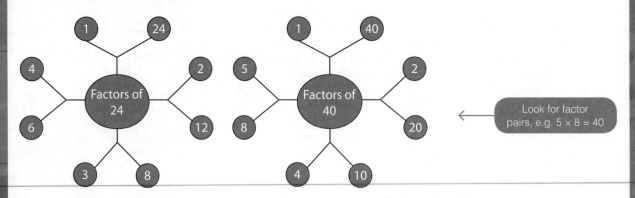

Look for factor pairs, e.g. $5 \times 8 = 40$

The factors of 24 are 1, 2, 3, 4, 6,⑧, 12 and 24.

Remember – every number has 1 and itself as a factor.

The factors of 40 are 1, 2, 4, 5,⑧, 10, 20 and 40.

8 is the HCF of 24 and 40.

Lowest common multiple

The lowest common multiple (LCM) is the lowest **multiple** shared by two or more numbers.

The first seven multiples of 5 are: 5, 10, 15, 20, 25,㉚,35, …

The first multiple of any number is itself.

The first seven multiples of 6 are: 6, 12, 18, 24,㉚,36, 42, …

30 is the LCM of 5 and 6.

Prime factors

Numbers can be expressed as the **product** of their **prime factors**. The product of the prime factors can be shown in **index form** where appropriate.

Write 24 as a product of its prime factors.

Use a factor tree to help. Start with the smallest **prime** number that divides exactly into 24 and continue in order until 1 is reached.

Prime numbers: 2, 3, 5, 7, 11, 13, … (1 is not a prime number).

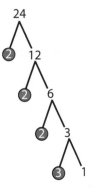

$24 = 2 \times 2 \times 2 \times 3$

In index form: $24 = 2^3 \times 3$

2

HCF and LCM using prime factors

You can find the HCF of two numbers by multiplying their shared prime factors.

You can find the LCM of the two numbers by multiplying the HCF with the rest of the prime factors.

Find the HCF and LCM of 24 and 42.

$24 = 2 \times 2 \times (2 \times 3)$

$42 = (2 \times 3) \times 7$

> This can be presented in a Venn diagram.

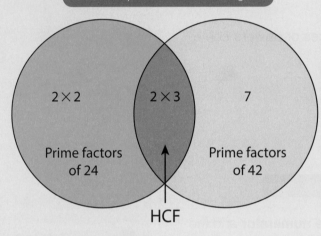

HCF

The **HCF** of 24 and 42 = 2 × 3 = **6**

The LCM of 24 and 42 = 6 × the rest of the prime factors

The **LCM** = 6 × 2 × 2 × 7 = **168**

Write the following prime numbers on separate pieces of paper.

| 2 | 2 | 3 | 3 | 5 | 7 |

Draw a Venn diagram on a piece of paper and place the prime numbers anywhere in the diagram (at least one in each of the three sections).

Find the product of each of the sections. You will now have two numbers and the HCF of those numbers. Find the LCM of your two numbers.

Repeat by rearranging your prime factors.

KEYWORDS

Factor ➤ A number which divides exactly into another number.

Multiple ➤ A number within the same times tables as another number.

Product ➤ The number or quantity obtained by multiplying two or more numbers together.

Prime factor ➤ A factor which is also prime.

Index form ➤ A number written using powers.

Prime ➤ A number with **exactly** two factors, 1 and itself.

1. (a) Find all the factors of 12 and 15.
   (b) Find the HCF of 12 and 15.
2. (a) List the first 8 multiples of 7 and 5.
   (b) Find the LCM of 7 and 5.
3. Answer these.
   (a) Write the following numbers as a product of their prime factors: 28 and 98.
   (b) Write your answers to part (a) in index form.
   (c) Find the HCF and the LCM of 28 and 98.

Using Bidmas

Bidmas helps you to remember the correct order of a calculation.

Brackets

Indices

Division

Multiplication

Addition

Subtraction

Start with brackets and then follow this order. Remember that multiplying and dividing comes before adding and subtracting.

Calculate $4 + 3 \times 6$ ← Multiply first and then add.

$4 + 18 = 22$

Remember that brackets followed by indices or **powers** come before any of the four operations.

Calculate $56 - 3^2(4 + 2)$

$56 - 3^2(6)$ ← Brackets

$56 - 9 \times 6$ ← Powers

$56 - 54 = 2$ ← Multiplication then subtraction

In the next example you must calculate the **numerator** and the **denominator** separately before dividing.

$$\frac{8 + 6}{(8 - 3) \times 20} = \frac{14}{5 \times 20} = \frac{14}{100} = 0.14$$

Reciprocals

The **reciprocal** of n is $1 \div n$ or $\frac{1}{n}$

Therefore the reciprocal of 5 is $1 \div 5$ or $\frac{1}{5}$

The reciprocal of $\frac{2}{3}$ is $1 \div \frac{2}{3} = 1 \times \frac{3}{2} = \frac{3}{2}$

To find the reciprocal you can simply flip the fraction upside down.

Adding and subtracting fractions

You can only add or subtract fractions which have a common denominator.

$\frac{1}{5} + \frac{3}{5} = \frac{1 + 3}{5} = \frac{4}{5}$ ← Keep the denominator the same.

If the denominators are different then find equivalent fractions with the same denominator.

$\frac{2}{3} - \frac{1}{4}$ ← The LCM of 3 and 4 is 12.

× by 4   × by 3

$\frac{8}{12} - \frac{3}{12}$ $= \frac{8 - 3}{12} = \frac{5}{12}$ ← Subtract as normal.

Operations

Module 3

Multiplying and dividing fractions

When multiplying fractions, simply multiply the numerators and then multiply the denominators.

$$\frac{2}{3} \times \frac{3}{5} = \frac{2 \times 3}{3 \times 5} = \frac{6}{15} = \frac{2}{5}$$ ← Don't forget to simplify.

When dividing fractions you don't actually divide at all. Find the reciprocal of the second fraction and then multiply instead.

multiply instead of divide

$$\frac{2}{3} \div \frac{4}{5} = \frac{2}{3} \times \frac{5}{4} = \frac{2 \times 5}{3 \times 4} = \frac{10}{12} = \frac{5}{6}$$

reciprocal of 2nd fraction

$$\frac{2}{3} \div \frac{4}{5} = \frac{5}{6}$$

Calculating with mixed numbers

When multiplying or dividing with mixed numbers, it's best to convert them to improper fractions.

$$2\frac{1}{4} \times 1\frac{2}{3} = \frac{9}{4} \times \frac{5}{3} = \frac{9 \times 5}{4 \times 3} = \frac{45}{12} = \frac{15}{4} = 3\frac{3}{4}$$

change to improper fractions

$$2\frac{1}{4} \times 1\frac{2}{3} = 3\frac{3}{4}$$

When adding or subtracting mixed numbers, you should deal with the fraction part separately.

$$2\frac{1}{4} \quad + \quad 3\frac{1}{3}$$

$$2 + 3 = 5 \qquad \frac{1}{4} + \frac{1}{3} = \frac{3}{12} + \frac{4}{12} = \frac{7}{12}$$

$$2\frac{1}{4} + 3\frac{1}{3} = 5\frac{7}{12}$$

Product rule for counting

The product rule for counting is used to find the total number of combinations possible for a given scenario, e.g. if there are A ways of doing Task 1 and B ways of doing Task 2, there are $A \times B$ ways of doing both tasks.

A man has five shirts, seven pairs of trousers and four jackets in his wardrobe.

To count the total combinations of different outfits he could wear, you can simply multiply:

$5 \times 7 \times 4 = 140$ different outfits

Fold a piece of A4 paper in half three times. Open it out and shade $\frac{3}{8}$ of it. Fold the paper back in half five times. Open it out and shade in $\frac{9}{32}$ (make sure you shade nine unshaded areas).

Count the total number of shaded squares to find the answer to $\frac{3}{8} + \frac{9}{32}$.

---

1. Calculate the following.
   (a) $6 + 3 \times 4$
   (b) $(7 - 4) \times (6 + 2)$
   (c) $(3 + 4^2) \times 2$
   (d) $\dfrac{5^2 + 3}{2 \times \sqrt{49}}$

2. Find the reciprocal of:
   (a) $\frac{2}{5}$
   (b) $\frac{1}{7}$
   (c) 5
   (d) 0.25

3. Work out:
   (a) $\frac{4}{5} - \frac{3}{4}$
   (b) $2\frac{1}{4} + 3\frac{3}{8}$
   (c) $\frac{4}{7} \div \frac{3}{5}$
   (d) $2\frac{2}{3} \times 1\frac{3}{4}$

Index form and roots

The index number tells you how many times the base number is multiplied.

Base number   Index

$$8^3 = 8 \times 8 \times 8 \quad \leftarrow \text{ Multiply 8 three times.}$$

The following table shows you how to calculate **roots**.

Roots	How to say it	Reasoning
$\sqrt{25} = 5$	Square root of 25 equals 5	Because $5 \times 5 = 25$
$\sqrt[3]{8} = 2$	Cube root of 8 equals 2	Because $2 \times 2 \times 2 = 8$
$\sqrt[4]{81} = 3$	Fourth root of 81 equals 3	Because $3 \times 3 \times 3 \times 3 = 81$

Estimate $\sqrt{60}$.

Since $\sqrt{49} = 7$ and $\sqrt{64} = 8$, $\sqrt{60}$ would be between 7 and 8.

$\sqrt{60} = 8$ (to the nearest whole number)

4

Index laws

Multiplying $\longrightarrow$ Add powers to give final index.

$4^2 \times 4^3 = 4^{2+3} = 4^5$ $\qquad$ $a^2 \times a^{-3} \times a^6 = a^{2-3+6} = a^5$ $\quad \leftarrow$ Normal rules apply for negative numbers.

Dividing $\longrightarrow$ Subtract powers to give final index.

$5^7 \div 5^3 = 5^{7-3} = 5^4$ $\qquad$ $b^9 \div b = b^{9-1} = b^8$ $\quad \leftarrow$ If no power is given, assume it is 1.

Raising one power to another power $\longrightarrow$ Multiply powers to give final index.

$(3^2)^4 = 3^{2 \times 4} = 3^8$ $\qquad$ $(c^2)^{-3} = c^{2 \times -3} = c^{-6}$

Zero, negative and fractional indices

These rules can be used to evaluate more difficult indices.

Rule	Example
$x^0 = 1$	$5^0 = 1$
$x^{-n} = \dfrac{1}{x^n}$	$6^{-2} = \dfrac{1}{6^2} = \dfrac{1}{36}$
$x^{\frac{1}{n}} = \sqrt[n]{x}$	$27^{\frac{1}{3}} = \sqrt[3]{27} = 3$
$x^{\frac{m}{n}} = (\sqrt[n]{x})^m$	$16^{\frac{3}{4}} = (\sqrt[4]{16})^3 = (2)^3 = 8$

Any non-zero number raised to the power of 0 equals 1.

Write down the reciprocal and change the sign of the index.

Also see page 21.

Fold a piece of paper in half six times and measure its thickness.

If it were possible, how many times would you need to fold it until it was thicker than 30cm? Or taller than your house? Or taller than the Eiffel Tower?

Open the piece of paper which was folded six times and count the rectangles. Consider why the total number of rectangles is $2^6 = 64$.

KEYWORDS

Roots ➤ Square roots, cube roots and so on.

Standard form

Standard form is a useful way of writing very large or very small numbers using powers of 10.
A number written in standard form looks like this:

$$A \times 10^n$$

A power of 10 where n is an integer.

A is between 1 and 10 (not including 10).

43 000 000 is written as 4.3×10^7

Use all the non-zero digits.

Count all the digits after the 4 (1 + 6 = 7).

0.000 638 is written as 6.38×10^{-4} ← The digits have moved four decimal places.

Calculating with standard form

When multiplying or dividing, deal with the powers separately.

$$(4 \times 10^6) \times (8 \times 10^{-3}) = (4 \times 8) \times (10^6 \times 10^{-3}) = 32 \times 10^3$$

Remember to put your answer back into standard form → $= 3.2 \times 10^4$

When adding or subtracting, first use ordinary numbers.

$$5.1 \times 10^5 + 2.4 \times 10^4 = 510\,000 + 24\,000 = 534\,000$$
$$= 5.34 \times 10^5$$

Know the following powers of 10:

$1000 = 10^3$

$100 = 10^2$

$10 = 10^1$

$1 = 10^0$

$0.1 = 10^{-1}$

$0.01 = 10^{-2}$

$0.001 = 10^{-3}$

Surds

A surd is a root which cannot be simplified further, e.g.
$\sqrt{2}, \sqrt{3}, \sqrt{5}$. Surds are irrational. $\sqrt{16}$ is not a surd since
$\sqrt{16} = 4$ and is not irrational. These rules apply to surds:

$$\sqrt{a} \times \sqrt{b} = \sqrt{a \times b} \quad \frac{\sqrt{a}}{\sqrt{b}} = \sqrt{\frac{a}{b}} \quad (\sqrt{a})^2 = a$$

Simplify a surd by taking out a square number factor.

$$\sqrt{12} = \sqrt{4 \times 3} = \sqrt{4} \times \sqrt{3} = 2\sqrt{3}$$ ← 4 is a square number factor of 12.

Expand brackets with surds.

$$\sqrt{3}(4 + \sqrt{3}) = 4 \times \sqrt{3} + (\sqrt{3})^2 = 4\sqrt{3} + 3$$

Rationalising the denominator

It is best if the denominator is not a surd (i.e. irrational).

To rationalise $\dfrac{2}{\sqrt{3}}$ you can multiply the numerator and
the denominator by $\sqrt{3}$.

$$\frac{2}{\sqrt{3}} \times \frac{\sqrt{3}}{\sqrt{3}} = \frac{2 \times \sqrt{3}}{(\sqrt{3})^2} = \frac{2\sqrt{3}}{3}$$

Multiplying by $\frac{\sqrt{3}}{\sqrt{3}}$ is the same as multiplying by 1.

To rationalise $\dfrac{1}{3 + \sqrt{3}}$ you multiply the numerator and the
denominator by $3 - \sqrt{3}$ (this is the denominator with the
sign between the two terms changed).

$$\frac{1}{3 + \sqrt{3}} \times \frac{3 - \sqrt{3}}{3 - \sqrt{3}} = \frac{3 - \sqrt{3}}{3^2 - (\sqrt{3})^2} = \frac{3 - \sqrt{3}}{9 - 3} = \frac{3 - \sqrt{3}}{6}$$

See 'Difference of two squares', page 23.

1. (a) Simplify the following,
   leaving your answer in
   index form.
   (i) $3^4 \times 3^5$ (ii) $4^8 \div 4$
   (iii) $\dfrac{6^2 \times 6^8}{6^3}$
   (b) Evaluate the following.
   (i) 7^0 (ii) 5^{-3}
   (iii) $8^{\frac{2}{3}}$

2. (a) Write the following in
   standard form.
   (i) 6000   (ii) 2300
   (iii) 0.006 78   (iv) 0.15
   (b) Calculate the following,
   leaving your answer in
   standard form.
   (i) $(3 \times 10^4) \times (6 \times 10^5)$
   (ii) $(8 \times 10^3) \div (4 \times 10^{-3})$
   (iii) $(5.1 \times 10^6) - (3.9 \times 10^5)$

3. Simplify the following.
   (a) $\sqrt{50}$ (b) $\sqrt{5}(4 - \sqrt{5})$
   (c) $(4 + \sqrt{7})(3 - \sqrt{7})$

Converting between fractions, decimals and percentages

Make sure you know these key fraction, percentage and decimal equivalences.

Fraction	Decimal	Percentage
$\frac{1}{3}$	$0.\dot{3}$	$33\frac{1}{3}\%$
$\frac{1}{4}$	0.25	25%
$\frac{2}{5}$	0.4	40%
$\frac{3}{10}$	0.3	30%
$\frac{5}{100}$	0.05	5%
$\frac{75}{100}$	0.75	75%
$\frac{3}{8}$	0.375	$37\frac{1}{2}\%$

← $1 \div 4 = 0.25$, $0.25 \times 100 = 25\%$

Fractions to decimals

Convert simple fractions into decimals by division.

$\frac{5}{8} = 5 \div 8 = 0.625$

$$8 \overline{)5 . {}^50 {}^20 {}^40} = 0.625$$

Use long division if no calculator.

Terminating decimals to fractions

A **terminating decimal** is a decimal number where the digits after the decimal point do not go on forever, e.g. 0.2, 0.35 and 0.875.

Convert decimals into fractions using $\frac{1}{10}$, $\frac{1}{100}$ and $\frac{1}{1000}$, etc.

$0.8 = \frac{8}{10} = \frac{4}{5}$

$0.48 = \frac{48}{100} = \frac{12}{25}$

Use HCF to cancel fractions to their simplest form.

$0.134 = \frac{134}{1000} = \frac{67}{500}$

5

Recurring decimals to fractions

Recurring decimals such as $0.\dot{4}$, $0.\dot{2}\dot{3}$ and $0.1\dot{3}\dot{6}$ can be converted to fractions.

Convert $0.\dot{3}\dot{5}$ to a fraction. ← $0.\dot{3}\dot{5} = 0.353\,535\,353\,535...$

$x = 0.353\,535\,3535...$ ← Label $0.\dot{3}\dot{5}$ as x.

$100x = 35.353\,535\,3535...$ ← Multiply by 100 since two digits recur.

$99x = 35$ ← Subtract x to remove the recurring decimals.

$x = \dfrac{35}{99}$ ← Divide both sides by 99.

Terminating decimal ➤
A decimal number in which the digits do not repeat forever.

Recurring decimal ➤
A decimal number in which the digits repeat forever after the decimal point.

Multiplying by a decimal

Look at this example.

2.23×4.9

It is best to ignore the decimals and multiply the whole numbers using a written method:

200	20	3	
8000	800	120	40
1800	180	27	9

```
      2 2 3
  ×     4 9
  2 0 0 7
  8 9 2 0
1 0 9 2 7
```

$223 \times 49 = 10\,927$

$2.23 \times 4.9 = 10.927$ ← A quick estimate will help place the decimal point ($2.23 \times 4.9 \approx 2 \times 5 = 10$)

Dividing by a decimal

It is best to change the number you are dividing by to a whole number.

$45 \div 0.3 = 450 \div 3 = 150$ ← Multiplying both numbers by 10 gives the exact same calculation.

$1.23 \div 0.15 = 123 \div 15 = 8.2$ ← Multiplying both numbers by 100.

Draw a 5 × 5 grid onto squared paper. Shade $\dfrac{7}{25}$ of the grid.

Draw an extra line in the centre of every row and column to turn it into a 10 × 10 grid.

Count the shaded squares to find out what $\dfrac{7}{25}$ is as a percentage and a decimal.

1. Convert the following fractions to decimals.
   (a) $\dfrac{3}{5}$ (b) $\dfrac{7}{8}$ (c) $\dfrac{2}{3}$

2. Convert the following decimals to fractions.
   (a) 0.36 (b) 0.248 (c) $0.\dot{2}\dot{7}$

3. Calculate the following.
   (a) 6.3×4.9 (b) $12 \div 0.4$ (c) $7.2 \div 0.5$

Decimal places

It is important to give an answer to an appropriate **degree of accuracy**. Remember that you should never round a number until you have the final answer of a calculation.

The table shows the usual degree of accuracy appropriate to the units given.

Units	Degree of accuracy
Centimetres (cm)	1 decimal place
Metres (m)	2 decimal places
Kilometres (km)	3 decimal places

> Rounding to two decimal places (2 d.p.) will leave you with two digits after the decimal point.

Round 3.476 32 to 2 d.p.

> Round up since the next digit on the right is 5 or more.

$3 . 4 7 | 6 3 2 = 3 . 4 8$ (to 2 d.p.)

> Round after the second digit.   Look at the next digit on the right.

Round 2.459 73 to 3 d.p.

> Since the fourth digit is a 7 and the third digit is a 9, round the second digit up by 1.

$2 . 4 5 9 | 7 3 = 2 . 4 6 0$ (to 3 d.p.)

> Round after the third digit.

Significant figures

Rounding to significant figures (s.f.) is useful for very large and very small numbers. The first significant figure of any number is the first digit which is not zero.

2 4 0 3 2 0   0 . 0 2 4 0 3 2 0

> The next digits (4, 0, 3 etc.) on the right are the second, third, fourth (etc.) significant figures.

> First s.f.

The table shows some examples of rounding to 1, 2 and 3 s.f. Round off in the usual way.

Example	1 s.f.	2 s.f.	3 s.f.
40 857	40 000	41 000	40 900
0.004 592	0.005	0.0046	0.004 59
12.93	10	13	12.9
1.987	2	2.0	1.99

KEYWORDS

Degree of accuracy ➤ What a number has been rounded to.

Upper bound ➤ The highest possible limit.

Lower bound ➤ The lowest possible limit.

Estimation

It is important to have an approximate idea of an answer before starting any calculation. To estimate a calculation it is best to round each number to 1 significant figure.

Estimate $\dfrac{3.4 \times 37}{0.21}$

$\dfrac{3.4 \times 37}{0.21} \approx \dfrac{3 \times 40}{0.2} = \dfrac{120}{0.2}$

> Multiply the numerator and the denominator by 10.

$= \dfrac{1200}{2} = 600$

6

Module 6   Approximations (side margin)

Upper and lower bounds

When a measurement is taken, it is usually rounded to an appropriate degree of accuracy:
➤ The **upper bound** is the maximum possible value the measurement could have been.
➤ The **lower bound** is the minimum possible value the measurement could have been.

The bounds are exactly half the degree of accuracy above and below the measurement.

This table shows some examples of upper and lower bounds.
You can write the error interval using inequalities as shown.

Measurement	Degree of accuracy	Lower bound	Upper bound	Error interval
23mm	Nearest mm	22.5mm	23.5mm	$22.5 \leqslant x < 23.5$
54.6cm	1 decimal place	54.55cm	54.65cm	$54.55 \leqslant x < 54.65$
4.32m	2 decimal places	4.315m	4.325m	$4.315 \leqslant x < 4.325$
34 000km	2 significant figures	33 500km	34 500km	$33 500 \leqslant x < 34 500$

The upper bound is actually 23.4$\dot{9}$.

We use the < symbol to mean 'up to but not including' 23.5

Calculating with upper and lower bounds

When calculating it is important to know whether to use the upper or lower bounds.

A field measures 20m wide and 30m long to the nearest metre. Find its maximum possible area.

The upper bound for the width is 20.5m.

Multiplying the two upper bounds will give the biggest possible area.

The upper bound for the length is 30.5m.

The maximum area for the field is 20.5 × 30.5 = 625.25m²

This table gives the correct use of bounds to find the maximum and minimum values:

Calculation	Minimum value	Maximum value
$a + b$	$a_{Lower} + b_{Lower}$	$a_{Upper} + b_{Upper}$
$a - b$	$a_{Lower} - b_{Upper}$	$a_{Upper} - b_{Lower}$
$a \times b$	$a_{Lower} \times b_{Lower}$	$a_{Upper} \times b_{Upper}$
$a \div b$	$a_{Lower} \div b_{Upper}$	$a_{Upper} \div b_{Lower}$

Use a tape measure to find the length and width of your bedroom to the nearest metre.

Calculate the maximum area of carpet needed for your room using upper bounds.

1. Round the following.
   (a) 2.3874 to 2 d.p.   (b) 4.609 92 to 3 d.p.   (c) 43 549 to 2 s.f.   (d) 0.004 034 9 to 3 s.f.
2. Estimate $\dfrac{28 \times 4.9}{0.18}$
3. Find the upper and lower bounds of these measurements, which have been rounded to the accuracy stated.
   (a) 23m (nearest m)   (b) 4.7kg (1 d.p.)   (c) 5.23km (3 s.f.)
4. Answer these.
   (a) A hall is measured as 18m by 24m to the nearest metre. What is the minimum area of the floor?
   (b) $a = 2.3$ and $b = 4.9$ are rounded to 1 d.p. Find the maximum value of $\dfrac{a}{b}$ correct to 2 d.p.

Calculator buttons

On a scientific calculator there are several buttons which are essential to know.

The table shows the button and its use.

Button	Use	Example
x^2	**Square**	$16^2 \longrightarrow$ [16] [x^2] [=] $\longrightarrow 256$
x^3	**Cube**	$7^3 \longrightarrow$ [7] [x^3] [=] $\longrightarrow 343$
y^x	Power	$3^6 \longrightarrow$ [3] [y^x] [6] [=] $\longrightarrow 729$
$\sqrt{}$	Square root	$\sqrt{60} \longrightarrow$ [$\sqrt{}$] [60] [=] $\longrightarrow 7.75$ (to 2 d.p.)
$\sqrt[3]{}$	Cube root	$\sqrt[3]{125} \longrightarrow$ [2nd F] [$\sqrt[3]{}$] [125] [=] $\longrightarrow 5$
$\sqrt[x]{}$	xth root	$\sqrt[5]{32} \longrightarrow$ [5] [2nd F] [$\sqrt[x]{}$] [32] [=] $\longrightarrow 2$
$a\frac{b}{c}$	Fraction	$\frac{8}{5} \longrightarrow$ [8] [$a\frac{b}{c}$] [5] [=] $\longrightarrow 1\frac{3}{5}$
Exp	Standard form	$3.4 \times 10^4 \longrightarrow$ [3.4] [Exp] [4] [=] $\longrightarrow 34\,000$

Some calculators use the button $\boxed{\frac{b}{c}}$

Some calculators use the button $\boxed{\times 10^x}$

Using brackets

Typing a large calculation into a calculator will usually require the use of brackets.

Calculate $\dfrac{23 + 4.5^2}{6.9 - 2.3}$

Type [(] [23] [+] [4.5] [x^2] [)] [÷] [(] [6.9] [−] [2.3] [)] [=] 9.402 173 913

> Do not round off unless directed to do so.

It can be beneficial to do a check by calculating the numerator and denominator separately.

$\dfrac{23 + 4.5^2}{6.9 - 2.3} = \dfrac{43.25}{4.6}$ ← Show this working in an exam.

$\qquad = 9.402\,173\,913$

7

KEYWORDS

Square ➤ Multiply a number by itself.

Cube ➤ Multiply a number by itself twice.

Module 7

Standard form

Standard form calculations can be performed on a calculator. However, brackets are essential to separate the numbers.

Calculate $(2.1 \times 10^3) \times (4.5 \times 10^8)$

Type $(\ \boxed{2.1}\ \boxed{\text{Exp}}\ \boxed{3}\ )\ \boxed{\times}\ (\ \boxed{4.5}\ \boxed{\text{Exp}}\ \boxed{8}\ )\ \boxed{=}$ 9.45×10^{11} ← The calculator will display this answer in standard form.

Use the $\boxed{(-)}$ button for negative indices.

Calculate $(3.6 \times 10^4) \div (2.5 \times 10^{-3})$

Type $(\ \boxed{3.6}\ \boxed{\text{Exp}}\ \boxed{4}\ )\ \boxed{\div}\ (\ \boxed{2.5}\ \boxed{\text{Exp}}\ \boxed{(-)}\ \boxed{3}\ )\ \boxed{=}$ $14\,400\,000$

$= 1.44 \times 10^7$ ← Change the answer back into standard form if required.

Make the number 16 by using the following calculator buttons.

| 3 | 4 | 5 | (|) | $\sqrt[3]{}$ | y^x | + | = |

You must only use each button once.

1. Calculate the following.
   (a) 2.3^3
   (b) 4^6
   (c) $\sqrt[3]{512}$
2. Calculate the following. Write down all the figures on your calculator display.
   $$\frac{8 + 2.5^2}{\sqrt{20.25}}$$
3. Calculate the following. Give your answer in standard form to 3 s.f.
   (a) $(2.6 \times 10^5) \times (3.8 \times 10^{-2})$
   (b) $(4.7 \times 10^8) \div (6.3 \times 10^{-7})$

Number Mind Map

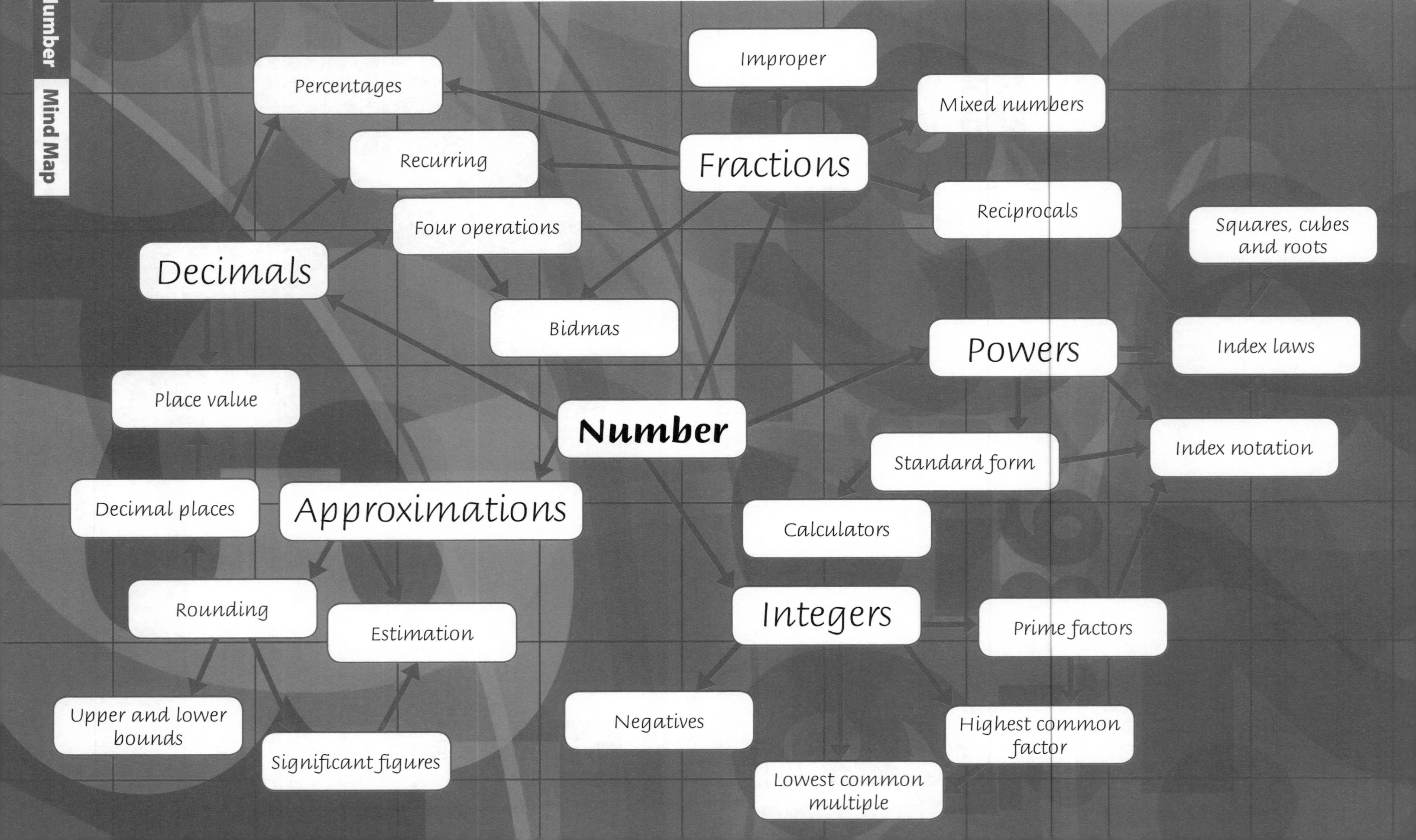

Number
Fractions
Improper
Mixed numbers
Reciprocals
Percentages
Recurring
Four operations
Bidmas
Decimals
Place value
Decimal places
Approximations
Rounding
Estimation
Upper and lower bounds
Significant figures
Powers
Squares, cubes and roots
Index laws
Index notation
Standard form
Calculators
Integers
Prime factors
Negatives
Highest common factor
Lowest common multiple

1. Work out the following. 🖩
    (a) 4×5.8 **[2]** (b) 2.3×42.7 **[3]**
    (c) $24 \div 0.08$ **[2]** (d) $46.8 \div 3.6$ **[2]**
    (e) $-3 + -4$ **[1]** (f) $-2 - -5 + -6$ **[1]**
    (g) $45 \div -9$ **[1]** (h) $-4 \times -5 \times -7$ **[1]**
    (i) $4 + 3^2 \times 7$ **[1]** (j) $(5 - 6^2) - (4 + \sqrt{25})$ **[2]**

2. (a) Write 36 as a product of its prime factors. Write your answer in index form. 🖩 **[2]**
    (b) Write 48 as a product of its prime factors. Write your answer in index form. 🖩 **[2]**
    (c) Find the highest common factor of 48 and 36. 🖩 **[2]**
    (d) Find the lowest common multiple of 48 and 36. 🖩 **[2]**

3. (a) Work out the following.
    $$\frac{4}{3} \times \frac{5}{6}$$
    Write your answer as a mixed number in its simplest form. 🖩 **[2]**
    (b) The sum of three mixed numbers is $7\frac{11}{12}$. Two of the numbers are $2\frac{3}{4}$ and $3\frac{5}{6}$.
    Find the third number and give your answer in its simplest form. 🖩 **[3]**
    (c) Calculate $3\frac{1}{5} \div 2\frac{1}{4}$.
    Give your answer as a mixed number in its simplest form. 🖩 **[3]**

4. Simplify the following. 🖩
    (a) $5^3 \div 5^{-5}$ **[1]** (b) $\sqrt{98}$ **[1]** (c) $\sqrt{7}(4 - 3\sqrt{7})$ **[2]**

5. Evaluate the following. 🖩
    (a) 2.3^0 **[1]** (b) $9^{-\frac{1}{2}}$ **[2]** (c) $27^{\frac{4}{3}}$ **[2]**

6. (a) Write the following numbers in standard form. 🖩
        (i) $43\,600$ **[1]** (ii) $0.008\,03$ **[1]**
    (b) Calculate the following, giving your answer in standard form. 🖩
        $(1.2 \times 10^8) \div (3 \times 10^4)$ **[2]**

7. Prove the following. You must show your full working out. 🖩
    (a) $0.4\dot{8} = \frac{16}{33}$ **[2]** (b) $0.1\dot{2}\dot{3} = \frac{61}{495}$ **[2]**

8. Rationalise the following surds. 🖩
    (a) $\frac{7}{\sqrt{5}}$ **[2]** (b) $\frac{3}{1 + \sqrt{2}}$ **[2]**

9. Paul's garage measures 6m in length to the nearest metre. His new car measures 5.5m in length to 1 decimal place.

    Is Paul's garage definitely long enough for his new car to fit in?
    Show your working. 🖩 **[3]**

10. $a = 4.3$ and $b = 2.6$ to 1 decimal place.

    Find the minimum value of $\frac{a}{b}$ and give your answer to 3 decimal places. **[3]**

Letters multiplied together

Make sure you know these rules:

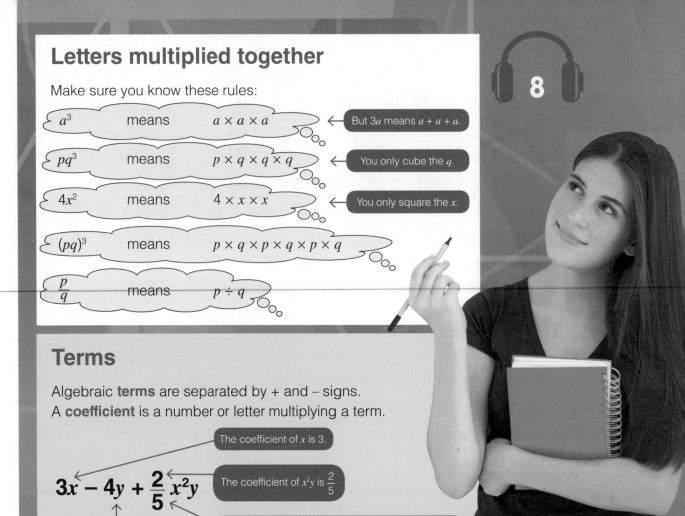

a^3 means $a \times a \times a$ ← But $3a$ means $a + a + a$.

pq^3 means $p \times q \times q \times q$ ← You only cube the q.

$4x^2$ means $4 \times x \times x$ ← You only square the x.

$(pq)^3$ means $p \times q \times p \times q \times p \times q$

$\dfrac{p}{q}$ means $p \div q$

Terms

Algebraic **terms** are separated by + and − signs.
A **coefficient** is a number or letter multiplying a term.

The coefficient of x is 3.

The coefficient of x^2y is $\dfrac{2}{5}$

$$3x - 4y + \frac{2}{5}x^2y$$

The coefficient of y is −4.

This term could be written as $0.4x^2y$ but it is more usual to use fractions as coefficients.

Simplifying expressions

An **expression** is a collection of (one or more) algebraic terms. 'Like terms' have the same combination of letters. Here is an example of an expression:

$9p$ and $-6p$ are like terms. They can be collected to give $3p$.

$$9p + 3pq + 5pq - 6p$$

$3pq$ and $5pq$ are like terms. They can be collected to give $8pq$.

So $9p + 3pq + 5pq - 6p = 3p + 8pq$

Combinations of letters that are **not** the same are said to be **unlike** terms.
For example, $2xy$ and xy^2 are unlike terms. They can't be added together.

KEYWORDS

Term ➤ A single number or variable, or numbers and variables multiplied together.

Coefficient ➤ The number occurring at the start of each term.

Expression ➤ A collection of terms.

Indices (powers) ➤ A convenient way of writing repetitive multiplication, e.g. $3^4 = 3 \times 3 \times 3 \times 3 = 81$. In this example, the power of 3 is 4.

Identity ➤ An equation that is true for all values.

Laws of indices

Make sure you know these seven rules:

Rule in words	Rule in symbols	Examples
When multiplying numbers with the same base, you add the **indices**	$a^m \times a^n = a^{m+n}$	$p^4 \times p^3 = p^{(4+3)} = p^7$ $2^7 \times 2^3 = 2^{10}$
When dividing numbers with the same base, you subtract the indices	$a^m \div a^n = a^{m-n}$	$q^8 \div q^3 = q^{(8-3)} = q^5$ $3^{13} \div 3^8 = 3^5$
When finding a power of a power, you multiply the powers	$(a^m)^n = a^{mn}$	$(a^2)^5 = a^{10}$ $(4^3)^4 = 4^{12}$
Any non-zero number to the power zero is 1	$a^0 = 1$	$q^0 = 1$ $8^0 = 1$
Negative powers can be made positive by using the reciprocal	$a^{-n} = \dfrac{1}{a^n}$	$a^{-3} = \dfrac{1}{a^3}$ $2^{-4} = \dfrac{1}{2^4} = \dfrac{1}{16}$
Fractional powers mean roots	$a^{\frac{1}{n}} = \sqrt[n]{a}$	$a^{\frac{1}{2}}$ means the square root of a, so $25^{\frac{1}{2}} = 5$ $a^{\frac{1}{3}}$ means the cube root of a, so $64^{\frac{1}{3}} = 4$
Fractional powers may be split into finding a root, followed by finding a power*	$a^{\frac{m}{n}} = \left(a^{\frac{1}{n}}\right)^m$	$125^{\frac{2}{3}} = \left(125^{\frac{1}{3}}\right)^2 = 5^2 = 25$

*Also remember – when finding the power of a **fraction**, you find the power of both the numerator and denominator of the fraction. So $\left(\dfrac{8}{27}\right)^{\frac{1}{3}} = \dfrac{8^{\frac{1}{3}}}{27^{\frac{1}{3}}} = \dfrac{\sqrt[3]{8}}{\sqrt[3]{27}} = \dfrac{2}{3}$

Equations and identities

An equation can be solved to find an unknown quantity. $5x + 3 = 23$ can be solved to find $x = 4$.

An **identity** is true for all values of x. $(5x + 3)^2$ is identically equal to $25x^2 + 30x + 9$. You can write this as $(5x + 3)^2 \equiv 25x^2 + 30x + 9$

Make nine cards with the letters $a, a, a, b, b, b, c, c, c$ on them. Turn all the cards face down and mix them up. Turn over two cards and multiply the letters, writing the answer algebraically, e.g. ab. Then repeat by picking three cards and multiplying the letters.

1. Simplify
   (a) $3ab - 2a^2b + 9ab + 4a^2b$
   (b) $2x - 7 + 7x - 2$
   (c) $ab + bc + ca + ac + cb + ba$

2. Simplify
   (a) $g^{12} \div g^4$
   (b) $\left(p^{15}\right)^{\frac{1}{3}}$
   (c) $2p^4q^{-3} \times 7p^{-2}q$

3. Evaluate
   (a) $216^{\frac{2}{3}}$
   (b) 4^{-2}
   (c) $\left(\dfrac{2}{3}\right)^{-3}$

Brackets

You need to be able to work with brackets:

Rule	Examples
To **multiply out** single brackets, multiply the term outside the brackets by every term inside.	$5(2a - 6) = (5 \times 2a) - (5 \times 6) = 10a - 30$ $3p(6p + 2q) = (3p \times 6p) + (3p \times 2q) = 18p^2 + 6pq$
To multiply out double brackets, multiply every term in the first bracket by every term in the second bracket.	$(2x + 3)(x - 6) = (2x \times x) + (2x \times -6) + (3 \times x) + (3 \times -6)$ $\qquad\qquad\qquad = 2x^2 - 12x + 3x - 18$ $\qquad\qquad\qquad = 2x^2 - 9x - 18$
When squaring brackets, always write out the expression in full to avoid making mistakes.	$(3x - 2)^2 = (3x - 2)(3x - 2) = 9x^2 - 6x - 6x + 4$ $\qquad\qquad\qquad\qquad\qquad\quad = 9x^2 - 12x + 4$ $\boxed{(3x - 2)^2 \text{ is } \textbf{not} \text{ equal to } 9x^2 + 4}$
When multiplying out three **binomials**, multiply out any two brackets to start with.	$(x + 1)(2x + 3)(x - 6) = (x + 1)(2x^2 - 9x - 18)$ Now multiply this with every term in the remaining bracket: $(x + 1)(2x^2 - 9x - 18) = x(2x^2 - 9x - 18) + 1(2x^2 - 9x - 18)$ $\qquad\qquad\qquad\qquad\qquad = 2x^3 - 9x^2 - 18x + 2x^2 - 9x - 18$ Finally, collect all like terms, giving: $(x + 1)(2x + 3)(x - 6) = 2x^3 - 7x^2 - 27x - 18$

Factorising

Factorising is the reverse of multiplying out brackets. Look for the largest number and the highest power that goes into each term.

$10x^2 - 5x = 5x(2x - 1)$ ← $5x$ is the largest factor of $10x^2$ and $5x$.

$6x^5 - 4x^2 = 2x^2(3x^3 - 2)$ ← 2 is the largest factor of both 6 and 4. x^2 is the highest power of x that goes into x^5 and x^2.

$18a^2b^6 - 12ab^4 = 6ab^4(3ab^2 - 2)$ ← $6ab^4$ is the largest factor of $18a^2b^6$ and $12ab^4$.

A quadratic expression is an expression of the form $ax^2 + bx + c$.

Factorising expressions where $a = 1$

$x^2 + 5x - 24$ ← Look for two numbers that multiply to give −24 and add to give 5. In this case $8 \times (-3) = -24$ and $8 + (-3) = 5$

So $x^2 + 5x - 24 = (x + 8)(x - 3)$

$x^2 + 18x + 65$ ← Look for two numbers that multiply to give 65 and add to give 18. The numbers required are 13 and 5.

So $x^2 + 18x + 65 = (x + 13)(x + 5)$

Always check you have the right answer by expanding the brackets: $(x + 13)(x + 5)$
$= x^2 + 13x + 5x + 65 = x^2 + 18x + 65$

Factorising expressions where $a \neq 1$

$2x^2 + 5x - 12$ ← The first two numbers in the brackets must be $2x$ and x since these are the only possibilities for a pair to multiply to give $2x^2$.

$2x^2 + 5x - 12 = (2x + ?)(x + ?)$

You now need two numbers that multiply to give −12, but they do **not** have to add up to 5.

Eventually you will find that −3 and 4 are the required pair, ← Possibilities are 12 and −1, −12 and 1, 2 and −6, 6 and −2 etc. In fact there are many possibilities. You have to try each pair by substituting them in and expanding the brackets.

i.e. $2x^2 + 5x - 12 = (2x - 3)(x + 4)$

Difference of two squares

Expressions with one square subtracted from another square can be factorised. These examples all follow the pattern $a^2 - b^2 \equiv (a + b)(a - b)$, an identity known as the **difference of two squares**.

$a^2 - 9 = a^2 - 3^2 = (a + 3)(a - 3)$

$16x^2 - 25y^6 = (4x)^2 - (5y^3)^2 = (4x + 5y^3)(4x - 5y^3)$

$3a^2 - 3 = 3(a^2 - 1) = 3(a + 1)(a - 1)$

Algebraic fractions

Simplifying algebraic fractions

Factorise the numerator and denominator, and cancel any common factors.

$$\frac{x^2 + x - 12}{2x^2 + 5x - 12} = \frac{(x + 4)(x - 3)}{(2x - 3)(x + 4)} = \frac{x - 3}{2x - 3}$$

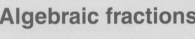

Here there is a common factor of $(x + 4)$

9

KEYWORDS

Multiplying out ➤ Removing brackets from an algebraic expression by multiplication.

Binomial ➤ An algebraic expression of the form $ax + b$, e.g. $x + 1$ or $2x - 3$.

Factorising ➤ Writing an algebraic expression as a product of factors, usually in brackets.

Difference of two squares ➤ The algebraic identity $a^2 - b^2 \equiv (a + b)(a - b)$

Algebraic fraction ➤ A fraction where the numerator, or denominator, or both, are algebraic expressions.

Adding (or subtracting) algebraic fractions

Find a common denominator. This can be found by multiplying the denominators together.

$$\frac{2}{x + 1} + \frac{3}{x + 2} = \frac{2(x + 2)}{(x + 1)(x + 2)} + \frac{3(x + 1)}{(x + 1)(x + 2)} = \frac{2x + 4 + 3x + 3}{(x + 1)(x + 2)} = \frac{5x + 7}{(x + 1)(x + 2)}$$

Multiplying algebraic fractions

Multiply the numerators and multiply the denominators. Always cancel down any common factors.

$$\frac{2}{x(x + 1)} \times \frac{x^2}{6(x + 2)} = \frac{2x^2}{6x(x + 1)(x + 2)} = \frac{x}{3(x + 1)(x + 2)}$$

There is no need to multiply out the $(x + 1)(x + 2)$ at the end.

Dividing algebraic fractions

Divide as you would for normal fractions. Change the divide sign to multiply, and take the reciprocal of the second fraction. Cancel any common factors.

$$\frac{x^2}{2x - 1} \div \frac{3x}{3x + 2} = \frac{x^2}{2x - 1} \times \frac{3x + 2}{3x} = \frac{x^2(3x + 2)}{3x(2x - 1)} = \frac{x(3x + 2)}{3(2x - 1)}$$

Write these six expanded expressions on separate pieces of paper and mix them up:

$x^2 + 6x + 9$ $\qquad$ $x^2 - 9$ $\qquad$ $x^2 - 6x - 9$

$x^2 - 6x + 9$ $\qquad$ $x^2 + 6x - 27$ $\qquad$ $x^2 - 9x$

Write these five factorised expressions on separate pieces of paper and mix them up:

$(x + 3)(x - 9)$ $\quad$ $x(x - 9)$ $\quad$ $(x + 3)^2$ $\quad$ $(x + 3)(x - 3)$ $\quad$ $(x - 3)^2$

Try to pair each expanded expression with the correct factorised expression. One of the expanded expressions cannot be factorised.

1. Multiply out
   (a) $5x(2 - 3x)$
   (b) $(2x + 5)(3x - 7)$

2. Factorise
   (a) $x^2 - x - 56$
   (b) $4p^2 - 25q^2$

3. Simplify
   (a) $\dfrac{x^2 - 8x + 16}{x^2 + 2x - 24}$
   (b) $\dfrac{1}{x + 2} + \dfrac{3}{x + 4}$

Using and rearranging formulae

Rearranging formulae is very similar to solving equations.

Make x the subject in the equation

$3x - y = z + x$

$2x - y = z$ ← Collect all x terms on one side of the equation.

$2x = z + y$ ← Put all other terms on the other side.

$x = \dfrac{z + y}{2}$ ← Divide through by 2.

Make x the subject in the equation

$\dfrac{5 + 12y}{x + 2y} = 8$

$5 + 12y = 8(x + 2y)$ ← Multiply through by $x + 2y$ to clear the fraction.

$5 + 12y = 8x + 16y$ ← Multiply out the brackets.

$5 - 4y = 8x$ ← Put all terms not including x on one side.

$x = \dfrac{5 - 4y}{8}$ ← Divide through by 8.

Make x the subject in the equation

$\dfrac{q - 3x}{4x + q} = q$ ← If the subject appears twice in the equation, you are likely to need to factorise once all the required terms are on one side of the equation.

$q - 3x = q(4x + q)$ ← Clear the fraction.

$q - 3x = 4qx + q^2$ ← Multiply out the brackets.

$q = 4qx + 3x + q^2$ ← Collect all x terms on one side.

$q - q^2 = 4qx + 3x$ ← Put all terms not including x on one side.

$q - q^2 = x(4q + 3)$ ← Factorise the terms involving x.

$x = \dfrac{q - q^2}{4q + 3}$ ← Divide through by $4q + 3$.

Make x the subject in the equation

$p + q = \sqrt{10x + y}$

$(p + q)^2 = 10x + y$ ← Square both sides to clear the square root sign.

$(p + q)^2 - y = 10x$ ← Put all terms not including x on one side.

$\dfrac{(p + q)^2 - y}{10} = x$ ← Divide through by 10.

Substituting numerical values in formulae

It is important to **substitute** numbers accurately into a given formula.

10

If $C = 3r^2$, find C when $r = 5$

$C = 3 \times 5^2 = 3 \times 25 = 75$ ← C is **not** $(3 \times 5)^2$. You only square the 5.

If $v = u + at$, find v when $u = 10$, $a = -2$ and $t = 8$

$v = 10 + (-2) \times 8 = 10 - 16 = -6$

If $v^2 = u^2 + 2as$, find v when $u = 5$, $a = -3$ and $s = 2$

$v^2 = 5^2 + 2 \times -3 \times 2 = 25 - 12 = 13$

So $v = \sqrt{13} = 3.61$

If $\dfrac{1}{f} = \dfrac{1}{u} + \dfrac{1}{v}$, find v when $f = 5$ and $u = 12$

$\dfrac{1}{5} = \dfrac{1}{12} + \dfrac{1}{v}$, so $\dfrac{1}{v} = \dfrac{1}{5} - \dfrac{1}{12}$ ← You could rearrange this formula to make v the subject first, though it is usually easier to substitute the numbers straight in.

$\dfrac{1}{v} = \dfrac{12}{60} - \dfrac{5}{60} = \dfrac{7}{60}$

So $v = \dfrac{60}{7} = 8.57$

Translating procedures into expressions or formulae

You may be asked to derive a formula from some given information.

A motorist wishes to rent a car. The rental cost is £35 per day, plus a standing charge of £60. Write down an expression for the total cost, £C, in terms of the number of days, d. For how many days can the motorist rent the car if he has £400 to spend?

The total cost is $C = 60 + 35d$

If $C = 400$, then you need to solve the equation $400 = 60 + 35d$

So $340 = 35d$ and $d = \dfrac{340}{35} = 9.71$

The motorist can rent the car for a maximum of 9 days.

The angles in a quadrilateral, in degrees, are $6y + 16$, $24 + y$, $106 - y$ and 130. Form an equation and use it to find the value of y.

The interior angles of a quadrilateral add up to $360°$, so $6y + 16 + 24 + y + 106 - y + 130 = 360$

Therefore $6y + 276 = 360$

So $6y = 84$ and $y = \dfrac{84}{6} = 14°$

S is the total sum of the interior angles in a regular polygon. If the polygon has n sides, find an expression for S in terms of n.

If the polygon has n sides, then it can be split up into $n - 2$ triangles.

For example, a regular six-sided hexagon can be split into four triangles.

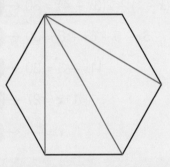

The interior angles of each triangle add up to $180°$, therefore $S = 180(n - 2)$

1. Make p the subject in each of the following formulae.

   (a) $\sqrt{p(a + 2)} = 5 - a$ (b) $q = \dfrac{3p + 2}{2p + 3}$

2. If $a = -10$, $b = 20$ and $c = 4$, find the exact values for p when

   (a) $p = a^2 - c^2$ (b) $p = \dfrac{a + c}{b}$

3. Joey cycles at a speed of s m/s.
   (a) If Alexia cycles 10m/s faster than Joey, find an expression in s for the time it takes Alexia to cycle 80m.
   (b) If Alexia cycles the 80m in 5 seconds, find Joey's speed.

The area of a trapezium is given by $A = \dfrac{(a + b)h}{2}$

Write each letter and symbol in this equation on separate pieces of card. You will also need a card with the minus symbol on it. Rearrange the letters and symbols to make a the subject of the equation. Then do the same to make b the subject and finally to make h the subject.

Solving linear equations

To solve any **linear equation**, you rearrange it to make x the subject.

> Remember to do the same operation to both sides of the equation.

Solve $\dfrac{x-1}{2} + \dfrac{3x+4}{5} = 3$

> If there are fractions in the equation, clear them by multiplying every term by the LCM of the denominators.

$\dfrac{10(x-1)}{2} + \dfrac{10(3x+4)}{5} = 30$

> The LCM of 2 and 5 is 10 so multiply every term by 10. Don't forget to multiply the term on the right-hand side.

$5(x-1) + 2(3x+4) = 30$

> Now cancel the fractions.

$5x - 5 + 6x + 8 = 30$

> Multiply out the brackets.

$11x + 3 = 30$

> Collect 'like terms'.

$11x = 27$

> Subtract 3 from both sides.

$x = \dfrac{27}{11}$

> Divide both sides by 11.

KEYWORDS

Linear equation ➤ An equation where each term is a multiple of one variable or a constant number.

Quadratic equation ➤ An equation where each term is a multiple of x^2, a multiple of x or a constant number.

Quadratic formula ➤ Allows you to solve all quadratic equations.

Completing the square ➤ Writing a quadratic expression

$$x^2 + px + q \text{ as } \left(x + \frac{p}{2}\right)^2 + r$$

Simultaneous equations ➤ A pair of equations in two unknowns.

Solving quadratic equations

Quadratic equations are usually written in the form $ax^2 + bx + c = 0$, where a, b and c are numbers.

Solving by factorising

> Factorising uses the idea that if two expressions multiplied together gives 0, then one of the expressions must be equal to 0.

Solve $3x^2 + 17x = 6$

$3x^2 + 17x - 6 = 0$

> Rearrange the equation so it is in the standard form.

$(3x - 1)(x + 6) = 0$

> Factorise.

So $3x - 1 = 0$ or $x + 6 = 0$

Therefore $x = \dfrac{1}{3}$ or $x = -6$

> Watch out for the 'difference of two squares'.

Solve $25x^2 - 144 = 0$

$(5x + 12)(5x - 12) = 0$

> Factorise.

So $5x + 12 = 0$ or $5x - 12 = 0$

Therefore $x = -\dfrac{12}{5}$ or $x = \dfrac{12}{5}$

Completing the square

Solve $x^2 + 4x - 7 = 0$ by **completing the square**.

$(x + 2)^2 - 11 = 0$

> 2 is half of 4.

> $(x + 2)^2 = x^2 + 4x + 4$ so you must subtract 11 to give -7.

$(x + 2)^2 = 11$

> Rearrange to make x the subject.

$x + 2 = \pm\sqrt{11}$

> Take the square roots of both sides.

$x = -2 \pm \sqrt{11}$

> Subtract 2 from both sides.

These are the answers in surd form. As decimals, the calculator will give $x = 1.32$ and $x = -5.32$

> Try to find different numbers a, b and c so that $b^2 = 4ac$.

Write down the solution(s) to the quadratic equation $ax^2 + bx + c = 0$ using your values of a, b and c.

Using the quadratic formula

The **quadratic formula** is given by

$$x = \frac{-b \pm \sqrt{b^2 - 4ac}}{2a}$$

Solve $x^2 - 9x + 7 = 0$

> Here $a = 1$, $b = -9$ and $c = 7$

So $x = \dfrac{-(-9) \pm \sqrt{(-9)^2 - 4 \times 1 \times 7}}{2 \times 1}$

So $x = \dfrac{9 \pm \sqrt{53}}{2}$

> b is already negative, so $-b = -(-9) = +9$

That is, $x = \dfrac{9 + \sqrt{53}}{2}$ and $x = \dfrac{9 - \sqrt{53}}{2}$

To 2 d.p. the solutions are therefore $x = 8.14$ and $x = 0.86$

Module 11 Algebraic Equations

Simultaneous linear and quadratic equations

Solve the **simultaneous equations** $\begin{aligned} 2x - 5y &= 23 \\ x + y &= 1 \end{aligned}$ ←

> The coefficients of x or those of y must be the same to **eliminate** them. Here neither are the same but the 2nd equation can be × by 2.

$2x - 5y = 23$
$2x + 2y = 2$ ←

> Now the coefficients of x are both 2, so **subtract** to eliminate x.

$-7y = 21$, so $y = -3$

$x + y = 1$ ←

> Substitute $y = -3$ back into either of the original equations.

$x - 3 = 1$, so $x = 4$

So the solutions are $x = 4$, $y = -3$

Solve the simultaneous equations $\begin{aligned} 4x - 3y &= 30 \\ 6x + 5y &= -31 \end{aligned}$

$20x - 15y = 150$
$18x + 15y = -93$

$38x = 57$, so $x = \dfrac{57}{38} = \dfrac{3}{2}$

> You can **eliminate** the x terms or the y terms. To **eliminate** the y terms, you can multiply the 1st equation by 5 and the 2nd equation by 3.

Substituting into the 1st equation: $6 - 3y = 30$

So $-3y = 24$ and $y = -8$

The solutions are $x = \dfrac{3}{2}$ and $y = -8$

> Now the signs of the coefficients of y are **not** the same, so **add** to eliminate y.

Solve the simultaneous equations $\begin{aligned} 2x^2 + 5y &= 23 \\ x + y &= 4 \end{aligned}$

Rearrange the 2nd equation to $y = 4 - x$ and **substitute** into the 1st equation to give
$2x^2 + 5(4 - x) = 23$, or
$2x^2 + 20 - 5x = 23$, or $2x^2 - 5x - 3 = 0$

$2x^2 - 5x - 3 = 0$ can be factorised to give
$(2x + 1)(x - 3) = 0$ ← So $2x + 1 = 0$ or $x - 3 = 0$

So solutions for x are $x = -\dfrac{1}{2}$ and $x = 3$

Substituting $x = -\dfrac{1}{2}$ back into the 2nd equation gives $-\dfrac{1}{2} + y = 4$, so $y = \dfrac{9}{2}$

Substituting $x = 3$ into the 2nd equation gives $3 + y = 4$, so $y = 1$

The solutions are $x = -\dfrac{1}{2}$, $y = \dfrac{9}{2}$ and $x = 3$, $y = 1$

Solving quadratic equations

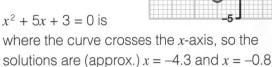

Graphs: This graph shows $y = x^2 + 5x + 3$ Use it to solve the equation
$x^2 + 5x + 3 = 0$

$x^2 + 5x + 3 = 0$ is where the curve crosses the x-axis, so the solutions are (approx.) $x = -4.3$ and $x = -0.8$

By drawing a straight line on the same grid you can solve $x^2 + 4x - 2 = 0$

Adding $x + 5$ to both sides of $x^2 + 4x - 2 = 0$ gives $x^2 + 5x + 3 = x + 5$ ←

> Start with the equation **you are trying to solve**, and add or subtract x^2 terms, x terms and numbers to obtain the equation **you are given**.

You already have the graph of $y = x^2 + 5x + 3$, so plot $y = x + 5$ and find the x-coordinates of the point(s) of intersection. You will find the solutions are (approx.) $x = -4.5$ and $x = 0.5$

Trial and improvement: Start by estimating a solution, then increasing it (if it is too small) or decreasing it (if it is too big) until you get as close as required.

Iterations: Solve the equation $x^3 + x = 12$ by iteration, giving your answer to 3 d.p.

$x^3 = 12 - x$, or $x = \sqrt[3]{12 - x}$ ←

> First rearrange the equation to make x the subject (it is likely you will already be given this in a rearranged form).

Now apply the iteration $x_{n+1} = \sqrt[3]{12 - x_n}$ to a value close to the final answer.

Call this estimate x_0.

Try $\quad x_0 = 2$

Then $\quad x_1 = \sqrt[3]{12 - 2} = 2.154...$ ←

And $\quad x_2 = \sqrt[3]{12 - 2.154...} = 2.143...$

$\quad\quad x_3 = \sqrt[3]{12 - 2.143...} = 2.144...$

$\quad\quad x_4 = \sqrt[3]{12 - 2.144...} = 2.144...$

> Do not round up at this stage! Use all the digits in the calculator's display in order to find x_2, then x_3 and so on.

Now you can say a solution is $x = 2.144$ to 3 d.p.

---

1. Solve $3x^2 - 11x - 20 = 0$ by factorisation.
2. Solve $x^2 - 12x + 18$ by completing the square. Leave your answers in surd form.
3. Solve the simultaneous equations $\begin{aligned} y &= 3x^2 - 9 \\ x + y &= 5 \end{aligned}$

Solving linear inequalities in one variable

The **inequality** symbols are:

< means less than	⩽ means less than or equal to
> means greater than	⩾ means greater than or equal to

Linear inequalities are solved in the same way as linear equations, with one main exception: when multiplying or dividing by a negative number, you change the direction of the inequality symbol.

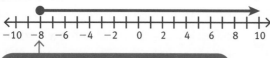
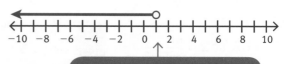

Solve the inequality $5x + 4 \geqslant 2x - 20$

$3x + 4 \geqslant -20$ ← Subtract $2x$ from both sides.

$3x \geqslant -24$ ← Subtract 4 from both sides.

$x \geqslant -8$ ← Divide by 3.

Solve the inequality $12 - 5x > 7$

$-5x > -5$ ← Subtract 12 from both sides.

$x < 1$ ← Both sides are divided by a negative number (–5) so the direction of the inequality is changed.

Both these solutions may be represented on a number line.

$x \geqslant -8$ is represented by a 'filled-in' circle (since –8 is included) and an arrow to the right.

$x < 1$ is represented by a 'hollow' circle and an arrow to the left.

These solutions may also be described using set notation or interval notation:

$x \geqslant -8$ may be written as $[-8, \infty)$ $x < 1$ may be written as $(-\infty, 1)$

The square bracket means the interval is 'closed' and contains the end point. The round bracket means the interval is 'open' and the end point is excluded. Infinity is not an actual number, so the infinity symbol is always written next to round brackets.

Solving and showing linear inequalities in two variables

You may be asked to shade a certain region on a graph that corresponds to two or more inequalities.

Shade the region corresponding to these inequalities: $x + y \geqslant 2$ $x < 2y + 2$ $y < 2$

$y = 2 - x$ $y = \frac{1}{2}x - 1$ $y = 2$ ← Change the inequality signs to = signs and rearrange each equation.

You can then draw a line for each equation. If the inequality is a 'strict inequality', i.e. if it is < or >, you should use dashed lines. Otherwise, use solid lines.

For $x + y \geqslant 2$, $0 + 0 \geqslant 2$ is **not** true so shade the other side of the *blue line*, i.e. the side **not** including the point (0, 0).

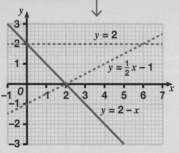
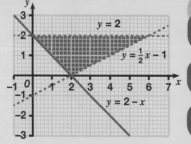

For $x < 2y + 2$, $0 < 2 \times 0 + 2$ **is** true so shade the side of the *red line* including (0, 0).

For $y < 2$, $0 < 2$ **is** also true so shade the side of the *green line* including (0, 0).

The shaded triangle now represents the intersection of all areas described above.

You need to decide which side of each line should be shaded. This can be done by using a simple 'test point'. (0, 0) is often the easiest test point to use.

Module 12   Algebraic Inequalities

Solving quadratic inequalities

To solve a **quadratic inequality**, the method is initially similar to solving quadratic equations. That is, make sure all the terms are on one side of the equation before trying to factorise.

Solve the inequality $x^2 + 3x < 10$

$x^2 + 3x - 10 < 0$ ← Rearrange.

$(x + 5)(x - 2) < 0$ ← Factorise.

You should now find the 'critical values'. These are where the left-hand side actually equals 0, i.e. when $x + 5 = 0$ or $x - 2 = 0$.

So $x = -5$ and $x = 2$ are the critical values.

Now sketch the graph of $y = (x + 5)(x - 2)$, knowing it crosses the x-axis at $x = -5$ and $x = 2$. It has a typical quadratic shape.

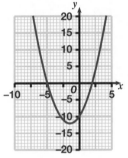

Now $y = (x + 5)(x - 2)$ is less than zero anywhere below the x-axis. That is, when x takes values between $x = -5$ and $x = 2$.

You can write the solution as $-5 < x < 2$ or in set notation as $(-5, 2)$.

The solution may also be represented on a number line:

1. Solve the inequality $2(x - 6) < 5(4 - 2x)$ and show your solution on a number line.
2. Solve the inequality $x^2 + 2x < 80$, writing your answer using set notation.
3. Shade the region corresponding to the following inequalities:
   $y < 4x - 3 \qquad y < 7 - x$
   $7y > 14 - 2x$

Chalk a number line from –5 to 5 on the ground. Practise showing these solutions to inequalities using filled-in circles, hollow circles and arrows:

$x \geqslant 1 \qquad x < 2 \qquad x > -4 \qquad x \leqslant 3$

Solve the quadratic inequality $2x^2 - 3x - 14 \geqslant 0$

$(2x - 7)(x + 2) \geqslant 0$ ← Factorise.

$2x - 7 = 0 \Rightarrow x = \dfrac{7}{2}$ ← Solve to find the critical values.

$x + 2 = 0 \Rightarrow x = -2$

Sketch the curve of $y = (2x - 7)(x + 2)$

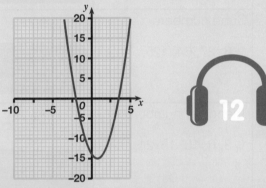

$y = (2x - 7)(x + 2)$ is greater than (or equal to) zero anywhere above the x-axis.

That is, when x takes values less than (or including) $x = -2$ or greater than (or including) $x = \dfrac{7}{2}$

The solution can be written as $x \leqslant -2$ or $x \geqslant \dfrac{7}{2}$

In set notation, this would be $(-\infty, -2] \cup \left[\dfrac{7}{2}, \infty\right)$

The $\cup$ symbol is a 'union' symbol and indicates both intervals are included in the solution.

On a number line:

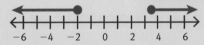

Inequality ➤ A relationship when one quantity is greater or less than another.

Linear inequality ➤ An inequality involving only x terms and numbers.

Quadratic inequality ➤ An inequality involving only x^2 terms, x terms and numbers.

KEYWORDS

Term-to-term and position-to-term sequences

A term-to-term **sequence** means you can find a rule for each **term** based on the previous term(s) in the sequence.

In the sequence 2, 6, 18, 54, 162, … you multiply by 3 to go from one term to the next.

> This can be written recursively as $U_{n+1} = 3U_n$ with $U_1 = 2$

The next term will be $162 \times 3 = 486$

In a position-to-term sequence, you can find a rule for each term based on its position in the sequence. In this example, the rule is $U_n = 2 \times 3^{n-1}$. In other words, the nth term in the sequence is $2 \times 3^{n-1}$

$U_1 = 2 \times 3^{1-1} = 2$ $U_2 = 2 \times 3^{2-1} = 6$
$U_3 = 2 \times 3^{3-1} = 18$ $U_4 = 2 \times 3^{4-1} = 54$

Standard sequences

Square numbers 1, 4, 9, 16, 25, …	$U_n = n^2$ This sequence is produced by squaring the numbers 1, 2, 3, 4, 5, … etc.	
Cube numbers 1, 8, 27, 64, 125, …	$U_n = n^3$ This sequence is produced by cubing the numbers 1, 2, 3, 4, 5, … etc.	
Triangular numbers 1, 3, 6, 10, 15, 21, …	$U_n = \dfrac{n(n+1)}{2}$ This sequence is generated from a pattern of dots that form a triangle.	
Fibonacci sequence 1, 1, 2, 3, 5, 8, 13, 21, …	$U_{n+2} = U_{n+1} + U_n$ and $U_1 = U_2 = 1$ The next number is found by adding the two previous numbers. So the next number in the sequence would be $13 + 21 = 34$.	There are other 'types' of Fibonacci sequence. One is the Lucas series: 2, 1, 3, 4, 7, 11, 18, … The term-to-term formula is the same as the original Fibonacci sequence, i.e. $U_{n+2} = U_{n+1} + U_n$, though in this case $U_1 = 2$ and $U_2 = 1$

The Fibonacci sequence is also a type of **recursive sequence** as finding the next term requires you to know previous terms.

Arithmetic sequences

An **arithmetic sequence** is a sequence of numbers having a common first difference.

Sequence   4 7 10 13 16   ← Here there is a common first difference of 3.

For any arithmetic sequence, the position-to-term formula is given by $U_n = dn + (a - d)$, where a is the first term and d is the common difference.

> The next number in the sequence will therefore be $U_6 = 3 \times 6 + 1 = 19$

So here $U_n = 3n + (4 - 3)$, i.e. $U_n = 3n + 1$

Module 13 Sequences

Geometric sequences

In a **geometric sequence** you multiply by a constant number r to go from one term to the next.

13

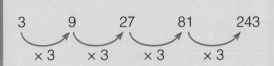

Each number is multiplied by 3 each time, so the common ratio $r = 3$

The term-to-term formula is $U_{n+1} = 3y_n$ with $U_1 = 3$

The position-to-term formula is $U_n = 3^n$

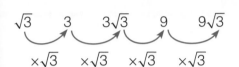

The common ratio $r = \sqrt{3}$

The term-to-term formula is $U_{n+1} = \sqrt{3}U$ with $U_1 = \sqrt{3}$

The position-to-term formula is $U_n = \left(\sqrt{3}\right)^n$

Finding the nth term of quadratic sequences

A **quadratic sequence** has position-to-term formula $U_n = an^2 + bn + c$, where a, b and c are constant numbers.

Find the first five terms in the quadratic sequence $U_n = 3n^2 - n + 2$

$U_1 = 3 \times 1^2 - 1 + 2 = 4$ $\qquad$ $U_2 = 3 \times 2^2 - 2 + 2 = 12$
$U_3 = 3 \times 3^2 - 3 + 2 = 26$ $\qquad$ $U_4 = 3 \times 4^2 - 4 + 2 = 46$
$U_5 = 3 \times 5^2 - 5 + 2 = 72$

The sequence is 4, 12, 26, 46, 72, …

You also need to be able to find the position-to-term formula for a quadratic sequence. This involves solving simultaneous equations to find the values of a, b and c.

Find the position-to-term formula, U_n, for the sequence 3, 4, 9, 18, 31, …

Sequence	3		4		9		18		31
First difference		1		5		9		13	
Second difference			4		4		4		

There is a constant second difference of 4, and so the sequence is quadratic, where $U_n = an^2 + bn + c$

Substitute: $n = 1$ $\qquad$ $U_1 = a + b + c = 3$ **1**

$\qquad$ $n = 2$ $\qquad$ $U_2 = 4a + 2b + c = 4$ **2**

$\qquad$ $n = 3$ $\qquad$ $U_3 = 9a + 3b + c = 9$ **3**

$\qquad\qquad\qquad$ $5a + b = 5$ **4** ← Equation 3 − Equation 2

$\qquad\qquad\qquad$ $3a + b = 1$ **5** ← Equation 2 − Equation 1

$\qquad\qquad\qquad$ $2a = 4$ so $a = 2$ ← Equation 4 − Equation 5

$\qquad\qquad\qquad$ $6 + b = 1$, so $b = -5$ ← Substitute $a = 2$ in Equation 5

$\qquad\qquad\qquad$ $2 - 5 + c = 3$ so $c = 6$ ← Substitute $a = 2$ and $b = -5$ in Equation 1

Therefore $U_n = 2n^2 - 5n + 6$

Draw three people in a triangle.

If every person shakes hand with each other, how many handshakes will there be?

Now draw four people in a square. Again, how many handshakes will there be?

Repeat for five people.

If n is the number of people, can you find a position-to-term formula for the number of handshakes?

How many handshakes are needed for 100 people?

1. The sequence 4, 13, 22, 31, 40, … forms an arithmetic sequence.
   (a) Find an expression for U_n, the position-to-term formula.
   (b) Hence find an exact value for U_{100}, the 100th term in the sequence.
   (c) Is 779 a number in this sequence? Explain your answer.
2. The sequence 5, 9, 15, 23, 33, … forms a quadratic sequence.
   (a) Write down the next term in the sequence.
   (b) If $U_n = an^2 + bn + c$, find the values of a, b and c.

Graphs of straight lines

To plot a straight-line graph, you should first draw up a table of values.

Plot the graph of $y = -2x + 4$

x	0	1	2	3
y	4	2	0	-2

Plot the points and draw the line.

Take some simple values for x, perhaps 0, 1, 2 and 3, and calculate y for each value. For example, when $x = 3$, $y = -2 \times 3 + 4 = -2$

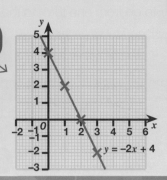

$y = -2x + 4$

$y = mx + c$

The equation of any straight line can be written as $y = mx + c$ where m is the **gradient** and c is the point of intercept with the y-axis. In the example above, the gradient m is -2 and the y-intercept, c, is 4. So the equation is $y = -2x + 4$

You need to be able to calculate both the gradient and the y-intercept from a given graph.

Find the equation of this straight line.

To find the gradient, take any two points on the line. Here we have chosen $(0, 3)$ and $(4, 5)$.

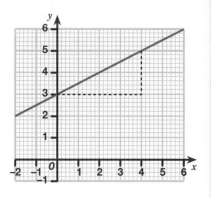

$$m = \frac{\text{change in } y}{\text{change in } x} = \frac{2}{4} = \frac{1}{2}$$

The graph crosses the y-axis at $(0, 3)$ so $c = 3$

Therefore, the equation of the line is $y = \frac{1}{2}x + 3$

Finding the equation of a line through two given points

Here is a line going through the points $(-2, 5)$ and $(4, 1)$.

The gradient

$$m = \frac{y\text{-step}}{x\text{-step}} = \frac{-4}{6} = -\frac{2}{3}$$

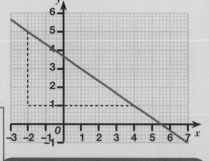

Now the equation of the line is $y = mx + c$,

or $y = -\frac{2}{3}x + c$

Notice from $(-2, 5)$ to $(4, 1)$, y steps down, so the y-step is negative. The gradient is negative since the line slopes downwards.

Substituting $x = 4$, $y = 1$ gives $1 = -\frac{8}{3} + c$,

so $c = \frac{11}{3}$

This can be written as $3y = -2x + 11$

The equation of the line is therefore $y = -\frac{2}{3}x + \frac{11}{3}$

KEYWORDS

Gradient ➤ The gradient, or slope, of a straight line shows how steep it is.

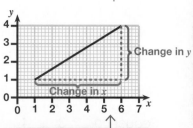

Change in y

Change in x

The gradient is equal to the change in y / change in x

Coordinates ➤ A pair of numbers (in brackets) used to determine a point in a plane.

Parallel ➤ Two lines are parallel if they have the same gradient. Parallel lines can never meet.

Perpendicular ➤ Two lines are perpendicular if they intersect at 90°.

Coordinates and Linear Functions

Module 14

Finding the equation of a line through one point with a given gradient

Find the equation of the line with gradient $\frac{1}{2}$, going through $(-3, 7)$.

The equation of the line is $y = mx + c$, or $y = \frac{1}{2}x + c$.

$7 = \frac{1}{2} \times (-3) + c$ ← Now substitute the **coordinates** of the point to find c, i.e. use $x = -3$ and $y = 7$

$7 = -\frac{3}{2} + c$

So $c = \frac{17}{2}$ and the equation of the line is $y = \frac{1}{2}x + \frac{17}{2}$ ← This can be written as $2y = x + 17$

Chalk two sets of x and y axes on the ground. On one set of axes, chalk a straight line going diagonally 'uphill'. On the second set of axes, chalk a straight line going diagonally 'downhill'. Pace out the changes in x and y to work out the rough gradient of each diagonal line.

Parallel and perpendicular lines

Two lines are **parallel** if they have the same gradient. For example, $y = 2x + 1$, $y = 2x$ and $y = 2x - 3$ are all parallel.

A line L is parallel to the line $y = \frac{1}{3}x + 4$ and passes through the point $(9, 2)$. Find the equation of L.

Since both lines are parallel, they both have the same gradient, and so L has gradient $m = \frac{1}{3}$

So L has equation $y = \frac{1}{3}x + c$

Substituting $x = 9$ and $y = 2$ gives $2 = \frac{9}{3} + c$, or $2 = 3 + c$.

So $c = -1$ and L has equation $y = \frac{1}{3}x - 1$

Two lines are **perpendicular** if they have gradients m and $-\frac{1}{m}$ ← This is the same as saying the product of their gradients is -1.

$y = -2x + 3$ and $y = \frac{1}{2}x - 2$ are perpendicular.

The first line has gradient -2 and the second has gradient $-\frac{1}{(-2)} = \frac{1}{2}$

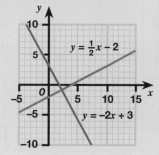

A line L is perpendicular to the line $y = 6x - 5$ and passes through the point $(18, 4)$. Find the equation of L.

Since both lines are perpendicular, L has gradient $m = -\frac{1}{6}$

So L has equation $y = -\frac{1}{6}x + c$

Substituting $x = 18$ and $y = 4$ gives $4 = -\frac{18}{6} + c$, or $4 = -3 + c$

So $c = 7$ and L has equation $y = -\frac{1}{6}x + 7$

1. Three lines are given by $4y = 16x - 4$, $4y = 8 - x$ and $4y = 8x + 16$ Which of the two lines are perpendicular?
2. Two perpendicular lines intersect at the point $(8, 0)$. If one line has gradient 4, find the equation of the other line.
3. Find the equation of the line joining points $(-2, -6)$ to $(6, 12)$.

Roots and intercepts

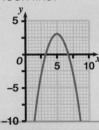

15

A quadratic function has the equation
$$y = ax^2 + bx + c$$

If $a > 0$, the graph will look like:

If $a < 0$, the graph will look like:

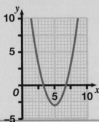

A quadratic function may also cross the x-axis twice (e.g. $y = x^2 - 6x + 5$), once (e.g. $y = x^2 - 4x + 4$) or not at all (e.g. $y = x^2 - 5x + 8$). However, it will always cross the y-axis at some point once.

Roots and **intercepts** of a quadratic function may be found graphically or algebraically.

This is the graph of
$$y = x^2 - 2x - 3$$

The graph crosses the x-axis twice, at $x = -1$ and $x = 3$. These are the roots of the equation $x^2 - 2x - 3 = 0$.
The intercept with the y-axis occurs at $(0, -3)$.

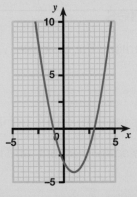

Consider the function $y = x^2 - 11x + 18$
To find the roots, factorise $x^2 - 11x + 18 = 0$ (or use the quadratic formula).
By factorising, you obtain $(x - 9)(x - 2) = 0$
$x - 9 = 0$ or $x - 2 = 0$
So the roots are $x = 9$ and $x = 2$

The curve crosses the y-axis when $x = 0$
So substituting $x = 0$ in the original equation:
$y = 0^2 - 11 \times 0 + 18 = 18$. Therefore, the graph will intercept the y-axis at $(0, 18)$.

Turning points

Every quadratic equation has a **turning point** at a minimum (if $a > 0$) or maximum point (if $a < 0$).

Finding a turning point graphically

The turning point on the curve
$y = x^2 - 2x - 3$
is a minimum point and occurs at $(1, -4)$.

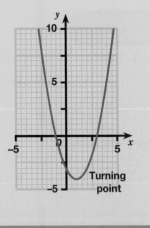

Turning point

Finding a turning point by completing the square

Find the minimum point on the curve of $y = 2x^2 - 3x + 8$ by completing the square.

$$y = 2\left[x^2 - \frac{3}{2}x + 4\right]$$ ← First factor out the 2.

$$y = 2\left[\left(x - \frac{3}{4}\right)^2 - \frac{9}{16} + 4\right]$$ ← Then complete the square.

$$y = 2\left[\left(x - \frac{3}{4}\right)^2 + \frac{55}{16}\right]$$ ← Then simplify.

$$y = 2\left(x - \frac{3}{4}\right)^2 + \frac{55}{8}$$ ← And multiply out.

Now $2\left(x - \frac{3}{4}\right)^2$ is always positive for any value of x, so its value is least when it is equal to zero.

That is, when $x = \frac{3}{4}$ ← When $x = \frac{3}{4}$, $y = \frac{55}{8}$

So minimum point on curve is $\left(\frac{3}{4}, \frac{55}{8}\right)$

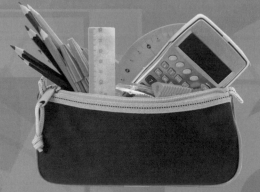

The symmetrical property of a quadratic

Solutions of the quadratic equation $ax^2 + bx + c = 0$

are given by $x = \dfrac{-b \pm \sqrt{b^2 - 4ac}}{2a}$

This can also be written as $x = \dfrac{-b}{2a} \pm \dfrac{\sqrt{b^2 - 4ac}}{2a}$

So the roots occur at an equal distance either side of the

line $x = -\dfrac{b}{2a}$

This is the **line of symmetry** of the quadratic function, and it can be shown that this is also the x-coordinate of the turning point on the curve.

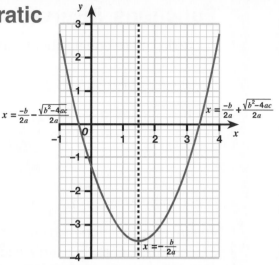

Finding a line of symmetry

Find the equation of the line of symmetry of the curve
$y = 3x^2 - 8x + 10$ ← $a = 3$ and $b = -8$

So line of symmetry is $x = -\dfrac{b}{2a} = -\left(\dfrac{-8}{6}\right) = \dfrac{4}{3}$

Finding a turning point by using the line of symmetry

Find the coordinates of the minimum point on the curve
$y = 2x^2 + 5x - 3$ ← $a = 2$ and $b = 5$

So x-coordinate of the minimum point is $x = -\dfrac{b}{2a} = -\dfrac{5}{4}$

The y-coordinate is $y = 2 \times \left(-\dfrac{5}{4}\right)^2 + 5 \times \left(-\dfrac{5}{4}\right) - 3$

$= \dfrac{25}{8} - \dfrac{25}{4} - 3 = -\dfrac{49}{8}$

So the minimum point has coordinates $\left(-\dfrac{5}{4}, -\dfrac{49}{8}\right)$

KEYWORDS

Roots ➤ In a quadratic equation $ax^2 + bx + c = 0$, the roots are the solutions to the equation.

Intercept ➤ The point(s) where a line or curve crosses another line, curve or axis.

Turning point ➤ In a quadratic curve, a turning point is the point where the curve has zero gradient. It could be a minimum or maximum point, and occurs at $x = -\dfrac{b}{2a}$

Line of symmetry ➤ For a quadratic curve $y = ax^2 + bx + c$, the line of symmetry is $x = -\dfrac{b}{2a}$. Both halves of the curve are identical either side of this line.

Draw a set of x-y axes on an A3 sheet of paper. Pick two specific points on the x-axis and one point on the y-axis. Draw a quadratic curve going through these points. Try to find the equation of the curve you have drawn. Is there more than one such equation?

1. Consider the curve $y = x^2 - 3x - 40$
   (a) Find the coordinates of the points where the curve intersects the x-axis and the y-axis.
   (b) Find the equation of the line of symmetry of the curve.
   (c) Hence find the coordinates of the turning point.
2. Consider the curve $y = x^2 + 10x - 40$
   (a) By completing the square, find the coordinates of the turning point.
   (b) Hence find the equation of the line of symmetry.

Functions, inverse functions and composite functions

A function maps one number to another number.

Consider the function 'add 3'.

1	maps to	4
2	maps to	5
3	maps to	6
4	maps to	7

> It is represented graphically by plotting $y = x + 3$

This can be written as $f(x) = x + 3$

The **inverse function** maps 4 back to 1, 5 back to 2, 6 back to 3 and 7 back to 4.

You write the inverse function as $f^{-1}(x)$. So in this case, $f^{-1}(x) = x - 3$

Composite functions consist of one or more functions.

Suppose $f(x)$ and $g(x)$ are two functions where $f(x) = x + 5$ and $g(x) = x^2$

The composite function $fg(x)$ means 'do g first, then f'.

> So $fg(10)$ for example, would be $f(g(10))$, or $f(100)$, which is 105.

So the composite function is $fg(x) = x^2 + 5$

As well as linear functions $f(x) = ax + b$ and quadratic functions $f(x) = ax^2 + bx + c$, there are other types of function you need to know.

Cubic and reciprocal functions

Cubic functions have the form

$$y = ax^3 + bx^2 + cx + d$$

> When $a > 0$, cubic functions generally have this shape:

The simplest cubic function is $y = x^3$

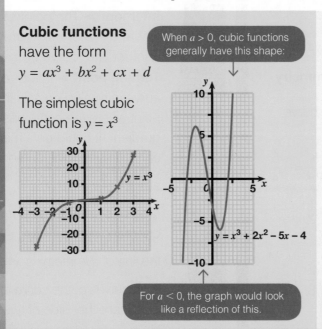

> For $a < 0$, the graph would look like a reflection of this.

Reciprocal functions have the form $y = \dfrac{k}{x}$ where k is a constant number. There are two types of curve, depending if k is + or –.

$y = \dfrac{2}{x}$　　$y = \dfrac{-3}{x}$

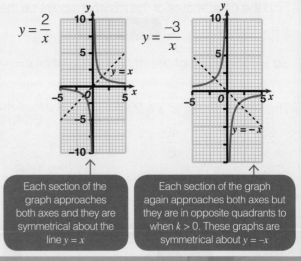

> Each section of the graph approaches both axes and they are symmetrical about the line $y = x$

> Each section of the graph again approaches both axes but they are in opposite quadrants to when $k > 0$. These graphs are symmetrical about $y = -x$

Exponential functions

Exponential functions have the form $y = a^x$ or $y = a^{-x}$, where a is a positive number. There are two types of curve but each type will always intersect the y-axis at (0, 1).

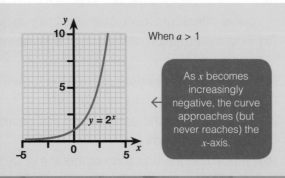

When $a > 1$

> As x becomes increasingly negative, the curve approaches (but never reaches) the x-axis.

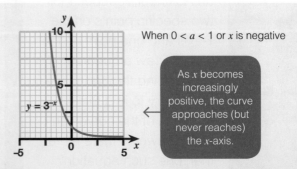

When $0 < a < 1$ or x is negative

> As x becomes increasingly positive, the curve approaches (but never reaches) the x-axis.

Trigonometric functions

sin x repeats itself every 360°.

cos x repeats itself every 360°.

$y = \tan x$ has asymptotes at $x = \pm 90°, \pm 270°, \pm 450°$ and so on. That is, the curve approaches these lines but never reaches them.

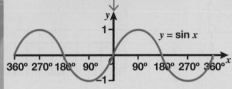

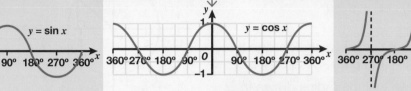

tan x repeats itself every 180°.

Transformations of functions

Suppose $f(x)$ is any function, and that a and k are positive constants. These examples show the effect of each transformation on $f(x) = \sin x$

16

$f(x) \rightarrow f(x) + a$ is a **translation** through a units in the direction of the y-axis.

$\sin x \rightarrow \sin x + 2$

This can also be described as a translation through the column vector $\binom{0}{a}$

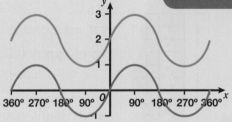

$f(x) \rightarrow f(x - a)$ is a translation through a units in the direction of the x-axis

$\sin x \rightarrow \sin(x - 90°)$

This can also be described as a translation through the column vector $\binom{a}{0}$

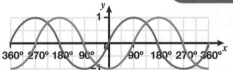

$f(x) \rightarrow -f(x)$ is a **reflection** in the x-axis.

$\sin x \rightarrow -\sin x$

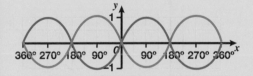

$f(x) \rightarrow f(-x)$ is a reflection in the y-axis.

$\sin x \rightarrow \sin(-x)$

In this example, a reflection of sin x in the x-axis gives the same result as a reflection in the y-axis, though this is not always the case with other functions.

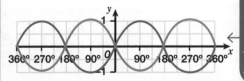

KEYWORDS

Translation ➤ A transformation where the whole graph is moved in a certain direction.

Reflection ➤ A transformation where the graph is 'reflected' in a certain line.

On A3 sheets of paper, draw two sets of axes from −360° to 360° on the x-axis and from −2 to 2 on the y-axis. Sketch the graphs of $y = \sin x$ and $y = \cos x$ on the axes using different coloured marker pens.

---

1. Sketch the graphs of $y = x^2$, $y = x^2 + 3$, $y = (x + 3)^2$ and $y = (-x)^2$
2. If $f(x) = 2x + 6$ and $g(x) = x^2 + 1$, find simplified expressions for $fg(x)$, $gf(x)$, $f^{-1}(x)$, $ff(x)$ and $gg(x)$.

Circles and tangents

The equation of a circle with radius r and centre $(0, 0)$ is $x^2 + y^2 = r^2$

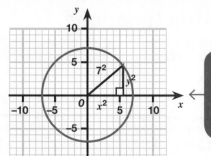

The radius of the circle is 7 units. By Pythagoras' theorem, $x^2 + y^2 = 7^2$ for every point on the circle, so the equation of the circle is $x^2 + y^2 = 49$

A **tangent** to a circle is a straight line that touches the circumference only once. It can be shown that the radius drawn to the **tangent line** is perpendicular to the tangent line. So if the radius has gradient m, then the tangent line has gradient $-\dfrac{1}{m}$.

The tangent line shown touches the circumference of the circle $x^2 + y^2 = 169$ at the point $(5, 12)$.

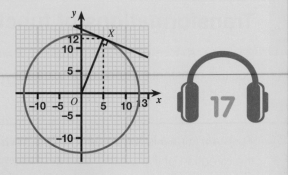

Find the equation of the tangent line.

The radius line has gradient $\dfrac{12}{5}$, therefore the tangent line has gradient $-\dfrac{5}{12}$, since it is perpendicular to OX.

The equation of the tangent line is $y = -\dfrac{5}{12}x + c$.

Substituting $x = 5$ and $y = 12$ gives $12 = -\dfrac{5}{12} \times 5 + c$

So $c = \dfrac{169}{12}$ and the equation of the tangent is

therefore $y = -\dfrac{5}{12}x + \dfrac{169}{12}$ ← This can be rearranged to $5x + 12y = 169$

If you are given a circle $x^2 + y^2 = r^2$ and a point (h, k) lying on the circumference, then the equation of the tangent line at (h, k) is $hx + ky = r^2$

KEYWORDS

Tangent line ➤ A line that touches a curve once, not crossing it.

Rate of appreciation ➤ The rate an object's value increases.

Rate of depreciation ➤ The rate an object's value decreases.

Gradients and areas under curves

The gradient of a curve is always changing. You can approximate the gradient at a point on the curve by drawing a tangent line at that point and finding the gradient of the tangent line.

The area under a curve can be approximated by splitting the curve into vertical 'strips' (each equivalent to a small trapezium) then adding them together. An example of this is shown on page 39.

Find an estimate for the equation of the tangent to the curve $y = x^3 - 3x^2 + 2$ at the point $x = \dfrac{5}{2}$

At $x = \dfrac{5}{2}$, $y = \left(\dfrac{5}{2}\right)^3 - 3 \times \left(\dfrac{5}{2}\right)^2 + 2 = -\dfrac{9}{8}$

Choose two points on the tangent line a distance apart, say $(4, 4.5)$ and $(0, -10.5)$.

Gradient of the line is approximately $\dfrac{-10.5 - 4.5}{-4} = 3.75$

Substituting $x = 0$ and $y = -10.5$ gives $-10.5 = 0 + c$, so $c = -10.5$

So equation of line is $y = 3.75x + c$

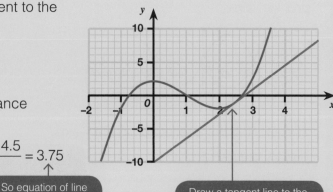

Draw a tangent line to the curve at $\left(\dfrac{5}{2}, -\dfrac{9}{8}\right)$

So equation of tangent line is $y = 3.75x - 10.5$

Distance–time and velocity–time graphs

In a **distance–time graph**, the gradient represents the **speed** at a given time. In a **velocity–time graph**, the gradient represents the **acceleration** at a given time. The area under a velocity–time graph represents the distance the object has travelled.

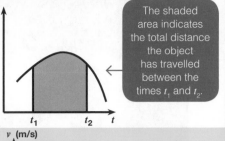

The shaded area indicates the total distance the object has travelled between the times t_1 and t_2.

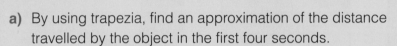

An object travels in a straight line from rest such that its velocity v at time t is given by $v = t \times 2^{-t}$ for $t \geqslant 0$. The graph of v against t is shown.

a) By using trapezia, find an approximation of the distance travelled by the object in the first four seconds.

When $t = 0$, $v = 0$

When $t = 1$, $v = 2^{-1} = \dfrac{1}{2}$

When $t = 2$, $v = 2 \times 2^{-2} = \dfrac{2}{2^2} = \dfrac{1}{2}$

When $t = 3$, $v = 3 \times 2^{-3} = \dfrac{3}{2^3} = \dfrac{3}{8}$

When $t = 4$, $v = 4 \times 2^{-4} = \dfrac{4}{2^4} = \dfrac{1}{4}$

Strip 1 has area $\dfrac{1}{2} \times 1 \times \dfrac{1}{2} = \dfrac{1}{4}$

Strip 2 has area $1 \times \dfrac{1}{2} = \dfrac{1}{2}$

Strip 3 has area $\dfrac{\frac{1}{2} + \frac{3}{8}}{2} \times 1 = \dfrac{7}{16}$

Strip 4 has area $\dfrac{\frac{3}{8} + \frac{1}{4}}{2} \times 1 = \dfrac{5}{16}$

The total distance travelled by the object is therefore $\dfrac{1}{4} + \dfrac{1}{2} + \dfrac{7}{16} + \dfrac{5}{16} = \dfrac{24}{16} = \dfrac{3}{2}$ m

b) Find the approximate time when the object's acceleration is 0m/s².

The acceleration in a velocity–time graph is given by the gradient, so you need to place a horizontal line on the graph in such a way that it is a tangent line.

You will find that the acceleration is 0m/s² after approximately 1.5 seconds.

Graphs and financial problems

Exponential graphs can help to estimate the **rate of appreciation** or the **rate of depreciation**.

A brand new car costs £20 000. Its value V after t years is given by $V = 20\,000 \times 0.6^t$

Estimate its rate of depreciation after 18 months.

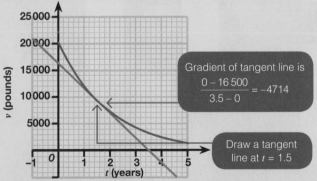

Gradient of tangent line is
$$\dfrac{0 - 16\,500}{3.5 - 0} = -4714$$

Draw a tangent line at $t = 1.5$

After 18 months, rate of depreciation is £4714 / year.

---

1. Sketch the circle of $x^2 + y^2 = 625$. Find the equation of the tangent to this circle at the point $(-7, 24)$.

2. Plot $y = \dfrac{1}{2}x^2$ for values of x from 0 to 4. Draw tangent lines to estimate the gradients at the points on the curve where $x = 1$ and $x = 3$. What do you notice?

On a big piece of paper, draw a circle with two tangent lines long enough to intersect. Comment on the distances from the point of intersection to the two places where they touch the circle.

Algebra

Inverse

Composite

Reflections

Number lines

Transformations

Inequalities

Functions

Translations

Set notation

Areas under curves

Quadratic formula

Estimating

Trial and improvement

Completing the square

Iteration

Algebra

Gradients

Quadratic

Simplifying

Factorising

Other equations

Expressions

Equations

Changing the subject

Simultaneous

Sequences

Tangents

Term-to-term

Circles

Position-to-term

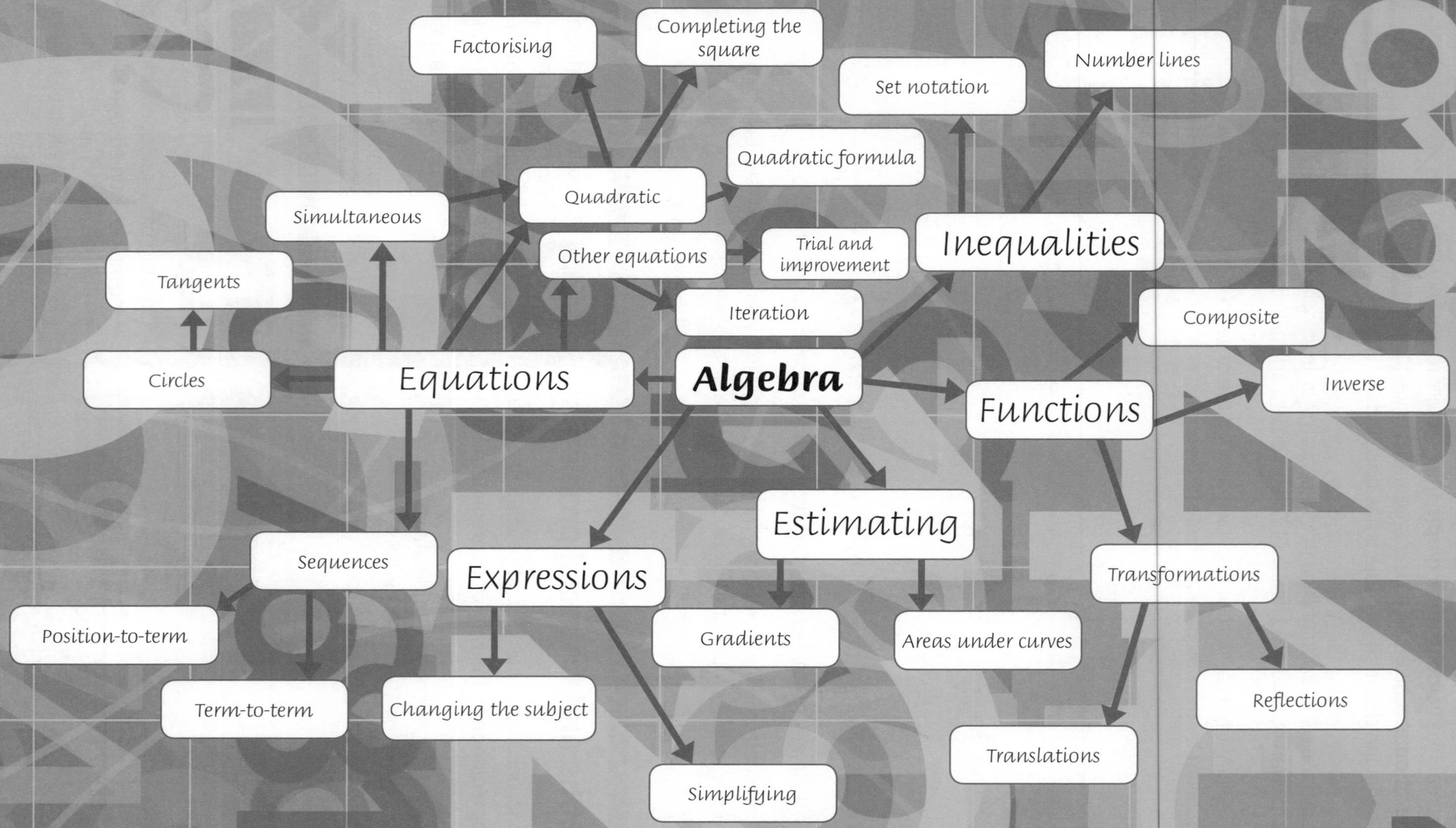

1. Simplify $\dfrac{5}{x-2} - \dfrac{1}{x-5}$ 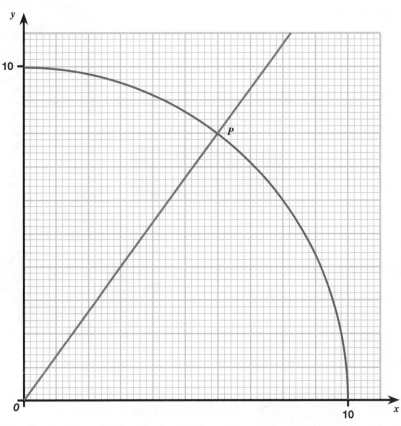 [2]

2. Which of the following lines are parallel and which are perpendicular? [2]

   A: $\boxed{3y = x - 6}$ B: $\boxed{4x + 2y = 8}$

   C: $\boxed{x = 3y - 12}$ D: $\boxed{2y - x + 2 = 0}$

3. The graph of $y = \cos x$ transforms to $y = 2 + \cos(x + 90)$ through a sequence of two transformations. Describe each transformation. [4]

4. (a) Solve $x^2 + 16x + 63 = 0$ by factorisation. [3]

   (b) Solve $x^2 - 8x + 10 = 0$ by completing the square. Give your answers in surd form. [3]

   (c) Solve $x^2 - 5x - 5 = 0$ by using the quadratic formula. Give your answers to 2 decimal places. [3]

5. If $y = \dfrac{3x + 1}{x - 2}$, make x the subject. [3]

6. Find the position-to-term formula, U_n, for the following sequence.

   5   21   47   83   129 [5]

7. The diagram shows a quarter-circle and a line with equation $y = \dfrac{4x}{3}$

(a) Find the equation of the circle. [2]

(b) Find the coordinates of P, the point of intersection of the line and the circle. [4]

(c) Find the equation of the tangent to the circle at P. [2]

Algebra

Metric and imperial units

In the UK we still use a mix of different units and you need to know some standard conversions between the two systems.

Length	2.5cm ≈ 1 inch	8km ≈ 5 miles
Volume	1 litre ≈ $1\frac{3}{4}$pts	4.5 litres ≈ 1 gallon
Mass	1kg ≈ 2.2lb	28g ≈ 1 ounce

Time

60s = 1 minute
60 mins = 1 hour
24h = 1 day
365 days = 1 year

Be very careful with time. Your calculator may give an answer of 2.25 hours. This is **not** 2h 25min but $2\frac{1}{4}$ h which is 2h 15min. If you have to enter 3h 30min into your calculator this is 3.5h **not** 3.30h.

The metric system

Metric measures are based on units of 10.

$milli...$ means $\frac{1}{1000}$	$centi...$ means $\frac{1}{100}$	$kilo...$ means 1000

Mass: 1000g = 1kg   and   1000kg = 1 tonne
Length: 1000mm = 100cm = 1m   and   1000m = 1km
Capacity: 1000ml = 1 litre   and   1000 litres = 1m³

Currency conversions

You must always use your common sense. Conversions are always given as a ratio, e.g. £1 : $1.60

Imagine the bank taking away your £1 coin and giving you $1.60 instead. You have more dollars than you had pounds (although they are worth the same).

Convert £450 to $ 450 × 1.6 = $720   ⟵ More dollars than pounds

Convert $620 to £   620 ÷ 1.6 = £387.50   ⟵ More dollars than pounds

Compound units

You need to know about speed (velocity), density and rates of pay.

Speed = $\frac{distance}{time}$ so the units will always be 'distance' per 'time'.

metres per second	centimetres per second	km per hour	miles per hour	feet per second
m/s or ms⁻¹	cm/s or cms⁻¹	km/h or kmh⁻¹	mph	ft/s or fts⁻¹

Density = $\frac{mass}{volume}$ so the units will always be 'mass' per 'volume'. ⟵

The most common density units are kg/m³ (or kgm⁻³) and g/cm³ (or gcm⁻³).

Rates of pay are given for a unit of time so £18/h or £500/week or £4400/month or £33000/annum.

The density of pure water is 1000 kgm⁻³ so anything which has a density less than this will float (including you at 985 kgm⁻³!)

Converting between **compound units** should always be done in steps.

Change 80km/h into m/s.

	80km	in 1 hour
is	80 000m	in 1 hour
is	80 000m	in 60min
is	1333.333...m	in 1min
is	22.22...m	in 1s

So 80km/h is 22.2m/s (3 s.f.)

Conversions for area and volume

A diagram always helps when converting between different units for area and volume.

Area

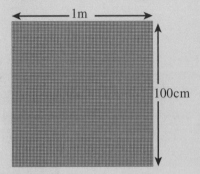

Filling a 1m² with cm²
- 100 rows of 100
- 10 000cm² (10^4 cm²) in 1m²

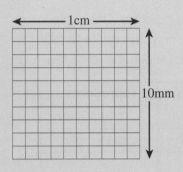

Filling a 1cm² with mm²
- 10 rows of 10
- 100mm² in 1cm²

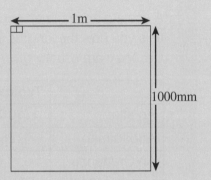

Filling a 1m² with mm²
- 1000 rows of 1000
- 1 000 000mm² (10^6 mm²) in 1m²

Volume

Imagine filling a box 1m × 1m × 1m with tiny cubes each 1cm × 1cm × 1cm. There would be 100 rows of 100 on the bottom layer and 100 layers. There are 1 000 000cm³ in 1m³.

Imagine filling a box 1cm × 1cm × 1cm with tiny cubes each 1mm × 1mm × 1mm. There would be 10 rows of 10 on the bottom layer and 10 layers. There are 1000mm³ in 1cm³.

Imagine filling a box 1m³ with mm³. There would be 1000 rows of 1000 and 1000 layers. There are 1 000 000 000mm³ (10^9) in 1m³.

KEYWORDS

Compound units ➤ A measurement unit using more than one quantity, often using 'per' as in speed, e.g. km/h (kilometres per hour).

Measure some water into a measuring jug. Weigh the jug of water and then weigh the jug empty. Work out the weight of the water and calculate its density in either gcm⁻³ or kgm⁻³.

1. Convert 64km/h to m/s.
2. Amil sees some jeans for sale in France for €80. If they cost £48 in the UK, where should he buy them? (£1 = €1.25)
3. Convert 8400mm² to cm².
4. Convert 2.5m³ to cm³.

Scale factors

You will meet scale factors in **scale drawings**, **similar triangles** and in **enlargements**.

A scale diagram is an accurate drawing. The scale tells you how the lengths in the diagram relate to real life. They are usually given as a ratio in the form 1: n

	Typical scale	...represents...
An atlas	1 : 10 000 000	1cm to 100km
A road map	1 : 250 000	1cm to 2.5km
An Ordnance Survey Explorer map	1 : 25 000	1cm to 250m
An architect's plan	1 : 100	1cm to 1m
A modeller's plan	1 : 24	1cm to 24cm

Scale factors and diagrams

If you have to calculate a distance from a map or plan:
➤ Measure the distance accurately with a ruler
➤ Write down the scale given
➤ Calculate the scaled-up 'real' size.

If you have to draw a map or plan:
➤ Decide on a suitable scale
➤ Write down the 'real' distances
➤ Calculate the scaled-down lengths
➤ Draw the plan accurately.

Beware!

The angles are always the same in both the plan and real life.

The areas on a scale drawing will be reduced by [scale factor]^2 – see Module 18.

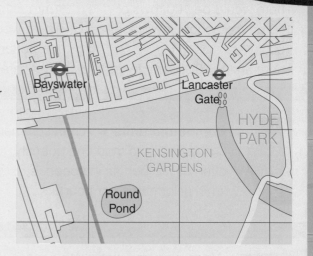

Scale 1 : 20 000

The distance between the Round Pond and Lancaster Gate station is 4cm.

In real life this is 800m.

Scales, Diagrams and Maps

KEYWORDS

Scale drawing ➤ A drawing used to represent a larger (or smaller) shape.

Similar triangles ➤ Triangles with corresponding angles equal and corresponding sides in the same ratio.

Enlargement ➤ A transformation that changes the size of an object using a scale factor and (usually) a centre of enlargement.

Bearings

Use bearings to set an absolute direction, measured from compass North. You always start at a point with your toes facing North. Then you turn clockwise until your toes are pointing in the direction that you want to go.

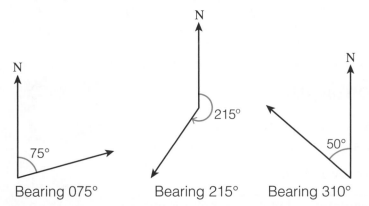

Bearing 075°   Bearing 215°   Bearing 310°

Use a compass to find North. Your school might have a 'bearings compass' or you might find one on a phone or digital watch. Mark this direction with a ruler or something similar and then try to find the bearing of some objects, for example the garden shed or a lamp-post.

If the bearing is less than 100° you insert a leading zero – all bearings must have three figures.

When you turn around and walk back the way you came, you turn through 180°.

If you have to reverse the direction of a bearing (a 'back bearing'), you simply add 180° to its value.

Bearing	15°	95°	110°	285°
Back bearing	15 + 180 = 195°	95 + 180 = 275°	110 + 180 = 290°	285 + 180 = 465° More than one complete turn so 465 – 360 = 105°

Bearings questions:
➤ can ask you to measure accurately or draw angles   (Module 24)
➤ can involve angles and parallel lines   (Module 25)
➤ can include Pythagoras' theorem and trigonometry.   (Module 32)

Ahab's sat-nav is broken but he has a compass and is sailing North. It is dark and he knows there are dangerous rocks. Coastguards A and B standing on the cliff can see his lights and measure his bearings as 160° and 230°.

Draw bearings and decide if he will hit the rocks if he carries on going North.

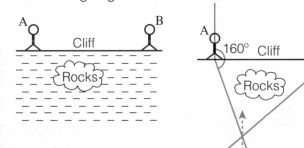

Ahab's position is where the two bearing lines cross. If he keeps going North, he will hit the rocks.

1. The scale of a road atlas is 1 : 250 000
   What do these distances represent in real life (in km)?
   (a) 4cm   (b) 10cm   (c) 1.5cm
   How far would these distances be on the map (in cm)?
   (d) 35km   (e) 40km   (f) 5km
2. Draw diagrams to show bearings of
   (a) 085°   (b) 190°   (c) 300°
   What is the back bearing in each case?

Fractions, decimals and percentages

Per cent means 'out of 100'. There is a direct link between fractions, decimals and percentages.

Fraction	$\dfrac{n}{100}$	$\dfrac{1}{100}$	$\dfrac{50}{100} = \dfrac{1}{2}$	$\dfrac{25}{100} = \dfrac{1}{4}$	$\dfrac{75}{100} = \dfrac{3}{4}$	$\dfrac{10}{100} = \dfrac{1}{10}$	$\dfrac{5}{100} = \dfrac{1}{20}$	$\dfrac{120}{100} = 1\frac{1}{5}$
Decimal		0.01	0.5	0.25	0.75	0.1	0.05	1.2
Percentage	$n\%$	1%	50%	25%	75%	10%	5%	120%

Calculating with percentages

There are many correct ways to calculate with percentages. The best method often depends whether you are allowed to use a calculator or not.

Three ways to calculate 'increase 480 by 21%' are shown.

Without a calculator	With a calculator	With a calculator and using a multiplier
1% of 480 = 4.8 10% of 480 = 48 20% of 480 = 96 480 + 96 + 4.8 = 580.8	21% of 480 = $\dfrac{21}{100} \times 480$ = 100.8 480 + 100.8 = 580.8	New amount is 121% of original. ← 'of' means '×' 1.21 × 480 = 580.8

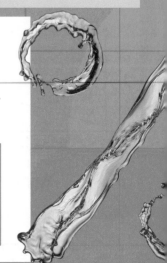

Comparisons using percentages

Percentages allow comparisons to be made easily as everything is written as parts of 100.

If you got 15 in a maths test, you can't tell how well you did unless you know the maximum possible marks. In each case write 15 as a fraction of the total available marks and then write it as a percentage.

15 out of 20 = $\dfrac{15}{20} = 0.75 = 75\%$

15 out of 30 = $\dfrac{15}{30} = 0.5 = 50\%$

15 out of 50 = $\dfrac{15}{50} = \dfrac{30}{100} = 30\%$ ← Equivalent fractions

Percentages and money

An article can **appreciate** (increase) or **depreciate** (decrease) in value by a given percentage.

➤ The value of a painting has appreciated by 15%. If its value **was** £45 000, it is now worth 1.15 × 45 000 = £51 750

➤ The value of a car has depreciated by 15%. If its value **was** £45 000, it is now worth 0.85 × 45 000 = £38 250
  ↑
  If it lost 15%, it would be now worth 85%

➤ If I bought a mansion for £2.5 million and sold it for £3.2 million: My **actual** profit would be £0.7 million or £700 000.
  My **percentage** profit would be
  $\dfrac{0.7}{2.5} = \dfrac{7}{25} = \dfrac{28}{100} = 28\%$

Simple and compound interest

When you invest a sum and then take the interest to spend (perhaps using the interest to live on), this is simple interest. Every year begins with the same sum invested to earn interest.

When you invest a sum and leave the interest in the account (to **compound**), each year the sum is larger and so the interest earned increases. It is easiest to calculate using a multiplier.

Simple	Compound	
£500 at 2% **per annum** Y1 500 × 0.02 = 10 Y2 500 × 0.02 = 10 Y3 500 × 0.02 = 10 Total interest £30	£500 at 2% per annum Y1 500 × 0.02 = 10 Y2 (500 + 10) × 0.02 = 10.20 Y3 (510 + 10.20) × 0.02 = 10.40 Total interest 10 + 10.20 + 10.40 = £30.60	Using a multiplier Y3 = 500 × 1.02³ = 530.60 Total interest 530.60 − 500 = £30.60

The compound interest principle also applies to repeated depreciation.

If a car worth £10 000 depreciates by 15% per year for five years, then its value is:

10 000 × 0.85⁵ = £4437   It loses 15% so 85% is left.

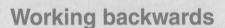

Make a drink of fruit cordial and measure how much concentrate and how much water you use. Then calculate the percentage of cordial in the drink.

Check the label on the bottle. Can you work out what percentage of your glass contains sugar?

Working backwards

The more difficult questions ask you to work backwards to find a value **before** an increase or decrease. You have to be very careful to read the question correctly, choose your method and show all your steps clearly.

20

A laptop costs £348.50 in a '15% off' sale. What was its price before the sale?

Using %	Using a multiplier
The sale price is 85% of the original. $\frac{348.50}{85} = 4.1$ (1% of the original). 4.1 × 100 = £410	The sale price is 85% of the original. 85% = 0.85 $\frac{348.50}{0.85} = 410$ so £410

A price increase of 8% means a new car is now £27 000. What was its price before the increase?

Using %	Using a multiplier
The new price is 108% of the original. $\frac{27000}{108} = 250$ (1% of the original). 250 × 100 = £25 000	The new price is 108% of the original. 108% = 1.08 $\frac{27000}{1.08} = 25000$ so £25 000

1. Without a calculator, work out these.
   (a) 15% of £56
   (b) Reduce 560 by 30%
   (c) The value of a £16 000 car after value added tax (VAT) of 20% is added.
2. Using a calculator, work out these.
   (a) 17% of £112
   (b) Increase 456 by 27%
   (c) 42 as a percentage of 520
3. If a plumber's bill was £460.80 after VAT at 20%, what was the price without VAT?

Writing ratios

You can use ratio, using colons (:), to show the proportion between two or more numbers.

If there are 25 dogs and 45 chickens, the ratio of dogs to chickens is
 25 : 45 or   ← Divide both parts by 5.
 5 : 9 in its simplest form.
But the ratio of dogs' feet to chickens' feet is
 100 : 90 or 10 : 9 in its simplest form.

If there are 25 dogs, 45 chickens and 20 sheep, the ratio of dogs : chickens : sheep is
 25 : 45 : 20 or
 5 : 9 : 4 in its simplest form.

Units should be the same when you make comparisons.
£2.50 : 50p as a ratio is 250p : 50p = 5 : 1
5m : 7000cm = 5m : 70m = 1 : 14

Unitary form (1 : n) is most commonly used in maps and scale drawings.
2 : 3 in unitary form is 1 : 1.5
4 : 5 in unitary form is 1 : 1.25
5 : 4 in unitary form is 1 : 0.8

Ratio and fractions

Reducing a ratio to its simplest form is similar to cancelling fractions.

16 : 18 simplifies to 8 : 9 ← Divide both parts by 2.

21 : 28 simplifies to 3 : 4 ← Divide both parts by 7.

There is a link between using ratio and fractions to divide amounts:

Ratio	1 : 2	2 : 3	3 : 5 : 8	1 : 1	1 : 4
Parts / Lots	3	5	16	2	5
Fractions	$\frac{1}{3}, \frac{2}{3}$	$\frac{2}{5}, \frac{3}{5}$	$\frac{3}{16}, \frac{5}{16}, \frac{8}{16}\left(=\frac{1}{2}\right)$	$\frac{1}{2}, \frac{1}{2}$	$\frac{1}{5}, \frac{4}{5}$

 Cut out 20 pictures of people, including both children and adults, from a newspaper or magazine. Organise your pictures into two groups – males and females – and write down the ratio of males to females in its simplest form. Then organise the pictures into children and adults and write down the ratio of children to adults in its simplest form.

Using ratio

To divide an amount in a given ratio, you first need to work out how many parts there are in the ratio.

Divide £840 in the ratio 6 : 5 : 1

There are 6 + 5 + 1 = 12 lots

$\frac{840}{12} = 70$ so 1 lot is £70

6 × 70 = 420; 5 × 70 = 350; 1 × 70 = 70

So £420 : £350 : £70 ← Check your answer by adding 420 + 350 + 70 = 840

Always read the question carefully.

Jane and Peter share a basket of apples in the ratio 3 : 5
If Peter had 15 apples, how many were there altogether?

Peter's apples are 5 parts of the ratio so
1 part is $\frac{15}{5}$ = 3 apples.

The whole basket was 8 parts, which is 3 × 8 = 24 apples.

There are different ways to compare the value of items.

Which is the best value?

A — 500g £3.50
B — 700g £4.75

For maximum marks, set out your working clearly.

Method 1 Find a price for equal amounts.

Jar A	Jar B
500g cost £3.50	700g cost £4.75
100g cost £0.70	100g cost £0.68

Jar B cheaper for 100g.

Method 2 Find how much you get for an equal price.

Jar A	Jar B
£3.50 buys 500g	£4.75 buys 700g
£1 buys 142.9g	£1 buys 147.4g

£1 buys more with Jar B so Jar B cheaper.

Method 3 Use common multiples.

Jar A	Jar B
500g cost £3.50	700g cost £4.75
× 7	× 5
3500g cost £24.50	3500g cost £23.75

Jar B cheaper for equal amounts.

1. Peter, Quinn and Robert have been left £44 800. They share it in the ratio of 3 : 5 : 6.
   How much does Quinn get? What fraction of the money does Robert get?
2. Which is the best value for money, 7.5m ribbon for £6.45 or 5.4m for £4.43?
3. Farmer Jones has 80 chickens. If the ratio of chicken's feet : sheep's feet is 2 : 3, how many sheep does he have?

Direct proportion

When two things are in a constant ratio, they are **directly proportional** (∝) to one another – as one increases so does the other.

There are many examples of direct proportion but the best is probably buying goods. The more bags of cement you buy, the more you will pay.

The total cost (C) is directly proportional to the number of items (n).

Inverse proportion

When two things are **inversely proportional** to one another, one goes up as the other goes down.

Sharing sweets is a good example of inverse proportion.

The more people who share, the fewer sweets they each get.

24 sweets to share: Two people get 12 each or three people get 8 each or twelve people get 2 each.

The number of sweets (s) is inversely proportional to the number of people (p).

22

Expressing proportions

Words	Symbols	Equation (k is a constant)
A is directly proportional to B	$A \propto B$	$A = kB$
A is directly proportional to B^2	$A \propto B^2$	$A = kB^2$
A is directly proportional to B^3	$A \propto B^3$	$A = kB^3$
A is directly proportional to $\sqrt{B}$	$A \propto \sqrt{B}$	$A = k\sqrt{B}$
A is inversely proportional to B	$A \propto \dfrac{1}{B}$	$A = \dfrac{k}{B}$
A is inversely proportional to B^2	$A \propto \dfrac{1}{B^2}$	$A = \dfrac{k}{B^2}$
A is inversely proportional to B^3	$A \propto \dfrac{1}{B^3}$	$A = \dfrac{k}{B^3}$
A is inversely proportional to $\sqrt{B}$	$A \propto \dfrac{1}{\sqrt{B}}$	$A = \dfrac{k}{\sqrt{B}}$

Solving proportion problems

Variables change in proportion in many areas of maths (and science) including problem-solving, similarity, transformations and trigonometry.

Problem-solving

Temperature (T) is inversely proportional to pressure (P). If T is 5.5 when P is 8, what is T when P is 4?

$$T = \frac{k}{P} \quad \text{so } 5.5 = \frac{k}{8}$$

$$\Rightarrow k = 5.5 \times 8 = 44$$

$$\Rightarrow T = \frac{44}{P}$$

When $P = 4$,

$$T = \frac{44}{4} = 11$$

KEYWORDS

$\propto$ ➤ The symbol for 'proportional to'.

Directly proportional ➤ Variables that are in a fixed ratio.

Inversely proportional ➤ As one variable increases, the other decreases.

Similar shapes ← Also see page 62.

A sheet of A4 paper, when measured to the nearest mm, is 210mm wide by 297mm long. A sheet of A5 is mathematically similar. If it is 148mm wide, what is its length?

Length $\propto$ width so $L = kW$

$$297 = k\,210 \Rightarrow k = \frac{297}{210} = \frac{99}{70}$$

$$\Rightarrow L = \frac{99}{70}W$$

When $W = 148$mm $L = \frac{99}{70} \times 148 = 209$ so 209mm

This problem could also have been solved using ratios.
If the shapes are similar then the ratio between corresponding sides is the same (one is an enlargement of the other).

$$210 : 297$$

$$= 1 : \frac{297}{210} \quad \leftarrow \text{Divide by 210 (leave as a fraction).}$$

$$= 148 : 209 \quad \leftarrow \text{Multiply by 148.}$$

Cut a sheet of A4 paper in half to make two pieces of A5 size. Cut one A5 piece in half again to make A6. How many A6 pieces would fit on an A4 sheet? If you can find A3 paper then cut that to make A4, A5 and A6. How many A6 sheets would fill one A3 sheet?

1. If 4m of wire costs £6.60, how much will 7m cost?
2. P is proportional to $\sqrt{Q}$ and P is 7 when Q is 9. Find the value of P when Q is 4.
3. C is inversely proportional to D^2. If $C = 3.5$ when $D = 1.5$, find D when $C = 5$

Rates of change and graphs

This graph shows the cost of a mobile phone on three different tariffs.

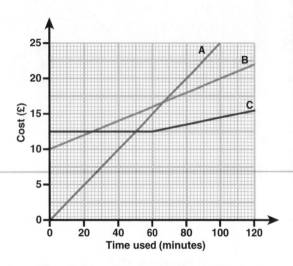

> Line A shows direct proportion. The **gradient** of the line is the rate of change.
$$\frac{\text{change in } y}{\text{change in } x} = \frac{10}{40} = 0.25$$
This is the constant of proportion and represents the cost per minute of 25p.

> Line B shows a fixed charge of £10 before any calls are made. The gradient is $\frac{5}{50} = 0.1$
This represents a cost per minute of 10p.

> Line C shows a fixed charge of £12.50 and all calls free up to a total of 60 minutes.
After that the gradient is $\frac{2.5}{50} = 0.05$
This represents a cost per minute of 5p for all calls after 60 minutes have been used.

You can use this graph to choose which is the best tariff for you. For more than 50 minutes use, C is cheaper, but for low use A is best.

Gradients of curves

If the graph is a curve then you can only estimate the gradient (rate of change) by drawing a **tangent** to the curve at the point you are interested in.

At the point (2, 2) draw a tangent line that just touches the line at the point but doesn't cross it.

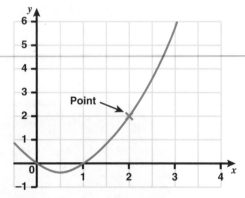

Then work out the gradient of that tangent.

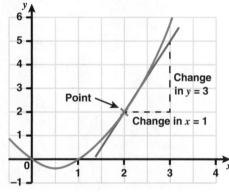

$\frac{3}{1} = 3$

At the point (2, 2) the gradient of the curve is 3 but it is different everywhere else.

From $x = 0.5$ upwards the rate of change (gradient) is increasing as the graph gets steeper and steeper.

Inverse proportion graphs

A graph showing inverse proportion looks like this – it is hyperbolic (a **hyperbola**).

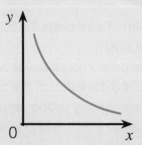

Take 25 counters (or make your own using pieces of paper). Increase the number of counters by 20%. Multiply the new number of counters by 40%. Reduce your latest number of counters by 50% and write down how many you are left with.

Rates of Change

Module 23

Repetitive rates of change

When a percentage change is applied more than once, you must take care to do each step at a time.

A new car costs £23 000. In the first year it depreciates by 20%, the next year it loses 15% and then 10% every year after that. To work out the value of the car after three years:

Step-wise
Lose 20% so worth 80%
80% of 23 000 = 0.8 × 23 000 = 18 400
then
85% of 18 400 = 0.85 × 18 400 = 15 640
then
90% of 15 640 = 0.9 × 15 640 = 14 076
Worth £14 076 after three years

Using multipliers
Lose 20% so worth 80%
Lose 15% so worth 85%
Lose 10% so worth 90%
0.8 × 0.85 × 0.9 = 0.612
0.612 × £23 000 = £14 076

23

1. Work out the gradient of these straight lines.

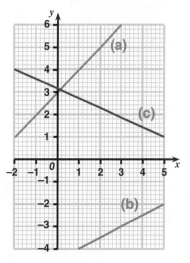

2. Draw a tangent to estimate the gradient of this curve at

   (a) $x = 2$ (b) $x = 3$ (c) $x = -1$

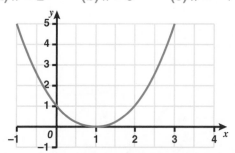

Ratio, Proportion and Rates of Change

Mind Map

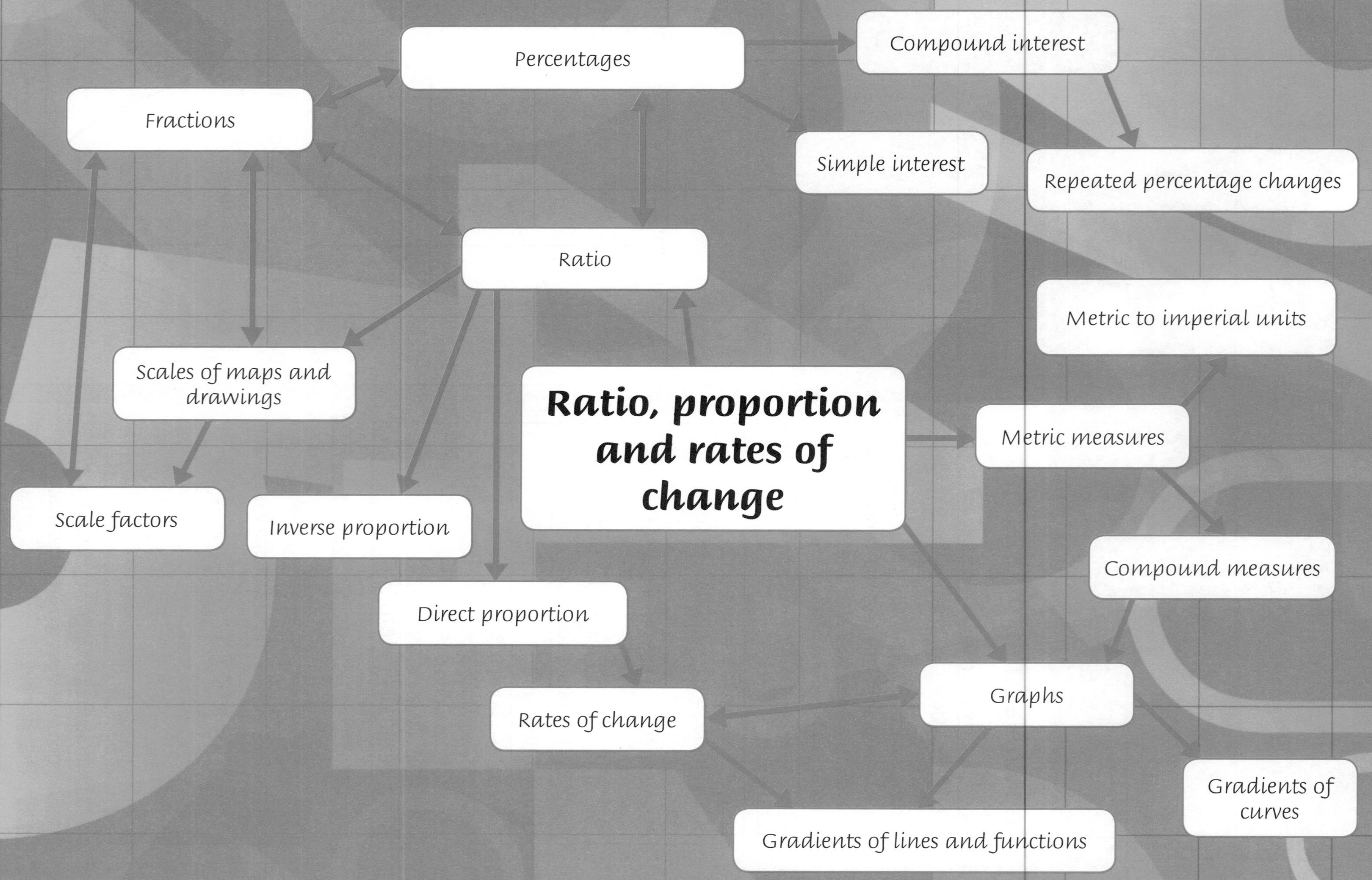

Ratio, proportion and rates of change

- Percentages
  - Compound interest
    - Repeated percentage changes
  - Simple interest
- Ratio
  - Metric measures
    - Metric to imperial units
    - Compound measures
  - Fractions
  - Scales of maps and drawings
    - Scale factors
  - Inverse proportion
  - Direct proportion
    - Rates of change
      - Graphs
        - Gradients of curves
        - Gradients of lines and functions

1. Convert 2.5m² to cm². 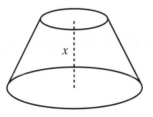 [2]

2. Alpha, Beta and Gamma share a legacy in the ratio 3 : 4 : 5

   If Beta had £2400, how much was left to them altogether?

   What fraction of the whole sum did Gamma inherit? [4]

3. Amil is buying a computer game on the Internet.

   If the exchange rate is £1 to €1.22, which should he buy? How much will he save?

[3]

4. James and Julia bought a house for £824 000 in 2009. It lost value and they sold it for £760 000 in 2013.

   What was their percentage loss on the transaction? [2]

5. A museum bought a painting for £45 000 in 2008. Museum staff calculate that its value went up by 4% in each of 2009, 2010 and 2011. In 2012 it lost 7% and in 2013 it lost a further 5%.

   (a) Was the painting worth more in 2014 than the museum originally paid? [3]
   (b) What single multiplier is equivalent to the five years of changes? [2]

6. S is inversely proportional to $\sqrt{M}$ and $S = 22.5$ when $M = 4$.

   What is the value of M when $S = 15$? [3]

7. The volume of the cone in the first diagram is 1600cm³.

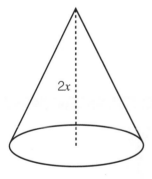

   If the top is cut off, as shown in the second diagram, what is the volume of the frustum that is left? [3]

Labelling and drawing

You must be able to draw lines of a given length accurately, and use a protractor to draw and measure angles to the nearest degree.

You will see triangles and angles labelled in many different ways.

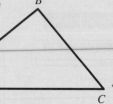

This triangle could be called triangle *ABC*, *BCA* or *CBA*.

The shaded angle could be correctly called the **acute angle** at *A* or just '*a*':

angle *BAC*   or   angle *CAB*

∠*BAC*   or   ∠*CAB*

BÂC   or   *CÂB*

← The side *CA* is extended to *D*.

The **obtuse angle** *DÂB* is an external angle.

Locus

Locus is a Latin word (plural **loci**) which means location or place. In geometry the locus of a set of points is all the places it can be, according to a given rule.

24

The locus of all points 3cm from the point *C*.	The locus of all points **equidistant** from the **line segment** *AB*.	The locus of all points equidistant from the points *P* and *Q*.	The locus of all points equidistant from the lines *FG* and *GH*.
A circle radius 3cm, centre *C* (or in 3D a sphere).	Two parallel lines joined with semicircles.	The **perpendicular bisector** of the line *PQ*.	The bisector of ∠*FGH*.
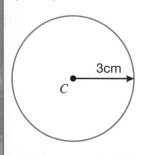	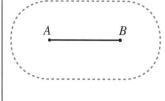	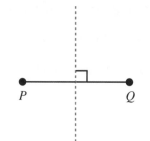	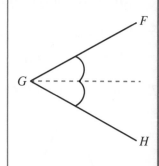

KEYWORDS

Acute angle ➤ An angle between 0° and 90°.

Obtuse angle ➤ An angle between 90° and 180°.

Equidistant ➤ The same distance from.

Line segment ➤ A short section of an infinite line.

Perpendicular to ➤ At 90° to.

Bisector ➤ Divides in half.

Locus/loci ➤ A set or sets of points.

Chalk a line segment on the ground. Using a tape measure or string to help you, chalk the locus of all the points 1 metre from the line segment.

Accurate constructions

You may have to construct angles and shapes without using a protractor to measure angles. You will only be allowed to use a ruler and a pair of compasses. ← Always leave all the construction arcs for the examiner to see.

You need to know that:

➤ a rhombus has diagonals that bisect its angles
➤ an isosceles triangle has a line of symmetry that bisects the base
➤ an equilateral triangle has all angles = 60°
➤ the shortest distance drawn from a point to a straight line is perpendicular to the line.

1. **The locus of all points equidistant from P and Q**

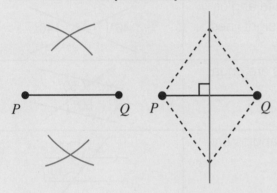

➤ You will construct two isosceles triangles with PQ as the base.
➤ Set compasses to a distance about the same as PQ.
➤ Use arcs to mark the vertices above and below PQ.
➤ Join the vertices – the line is the perpendicular bisector of PQ.

2. **The locus of all points equidistant from the lines FG and GH**

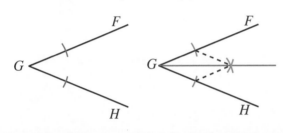

➤ You will draw a line that bisects the angle by constructing a rhombus.
➤ Set compasses to a fixed distance.
➤ From G mark two equal lengths on the lines FG and GH.
➤ Using these marks set the fourth vertex of the rhombus.
➤ Draw the diagonal. This is the angle bisector.

3. **Angles of 60° and 30°**
➤ Using a pair of compasses and a ruler, construct **any** equilateral triangle. This gives an angle of 60° **without** using a protractor.
➤ To draw 30° you need to bisect the 60° angle. Construct a rhombus – see **2**.

4. **Angles of 90° and 45°**
➤ Using a pair of compasses, construct a line and its perpendicular bisector to make 90° – see **1**.
➤ To draw 45° you need to bisect the 90° angle. Construct a rhombus – see **2**.

1. Draw the following:
   (a) The locus of all points 3cm from a line of length 5cm.
   (b) The locus of all points equidistant from points S and T which are 6cm apart.
   (c) Two intersecting lines AB and BC and the locus of all points equidistant from both lines.
2. Construct an angle of 30° using a pair of compasses and a ruler only.

25

Angle rules

Look at these angle rules and examples:

Angles at a point	The angles around a point add up to 360°.	107° 32° 221°
Angles on a straight line	The angles along a straight line add up to 180°.	82° 29° 69°
Vertically opposite angles	Vertically opposite angles between two straight lines are equal.	46° 46°
Parallel lines and alternate angles	Alternate angles are equal.	50° 50°
Parallel lines and corresponding angles	Corresponding angles are equal.	130° 130°
Angles in a triangle	Angles in a triangle always add up to 180°.	43° 92° 45°

The properties of triangles

The shape of a triangle is described using its angles and lengths.

Name	Sides	Angles	
Equilateral	Three sides equal; a **regular** triangle	Three angles equal (60°)	Has three lines of symmetry
Isosceles	Two sides equal	Two angles equal	Has one line of symmetry
Scalene	No sides equal	No angles equal	
Right-angled	Could be isosceles or scalene	One right angle	
Obtuse-angled	Could be isosceles or scalene	One **obtuse** angle	

Proof of angle sum for any triangle

You can use the angle rules to **prove** that the angles inside a triangle always add up to 180°.

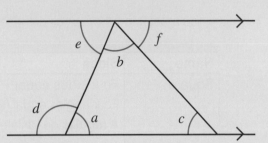

$e + b + f = 180°$ (Angles on a straight line sum to 180°.)
angle e = angle a (Alternate angles are equal.)
angle f = angle c (Alternate angles are equal.)
So $a + b + c = 180°$
Angles inside a triangle always add up to 180°.

Cut any triangle out of a sheet of paper. Colour in the angles at the corners and then tear the triangle into three. You will find that you can place the three angles together along a straight edge (they add up to 180°).

1. Find the missing angles, giving your reasons.

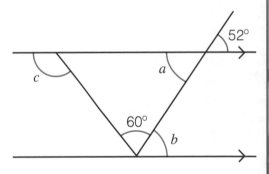

2. Find the angle s.

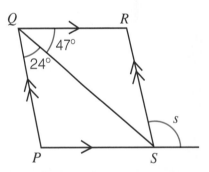

3. Find the angles x, y and z.

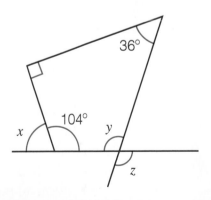

26

The properties of quadrilaterals

You must know the names of these special quadrilaterals. Quad – four, lateral – side

Name	Sides	Symmetry	
Square	Four sides equal Four angles equal (90°) A **regular** quadrilateral	Four lines Rotational order 4	
Rectangle	Two pairs of opposite sides equal Four angles equal (90°)	Two lines Rotational order 2	
Parallelogram	Two pairs of opposite sides equal and parallel (a squashed rectangle)	No lines Rotational order 2	
Rhombus	Four sides equal Two pairs of opposite sides parallel (a squashed square)	Two lines Rotational order 2	
Trapezium	One pair of parallel sides Could also be isosceles	No lines or one line Rotational order 1	
Kite (also a delta or arrowhead)	Two pairs of adjacent sides equal	One line Rotational order 1	

Make any large polygon shape on the floor – you could use metre rules or just mark the corners with books. Start at one corner and walk round your shape, stopping at each corner and turning. This turn is an external angle. When you get back to your start point turn, facing exactly the same way as you started. You will have turned through 360°. The external angles of a polygon add up to 360°, however many sides it has.

Properties of 2D Shapes

Module 26

The properties of all polygons

Polygons are named by the number of sides.

Sides	3	4	5	6	7	8	9	10
Name	Triangle	Quadrilateral	Pentagon	Hexagon	Heptagon	Octagon	Nonagon	Decagon

A regular polygon has all sides and angles equal. ← A square is a regular quadrilateral.

Polygons have both **external** angles and **internal** angles.

The external angle of any polygon is the angle between one side extended and the next side.

The external angles of any polygon always add up to 360°.

There are two common methods to work out the internal angles in a polygon:

Method 1

➤ Assume the polygon is regular and work out the size of each external angle.

➤ Use the 'angles on a straight line' rule to work out each internal angle.

➤ Calculate the total sum of all the internal angles.

 72°

 108°

 540°

$\frac{360}{5} = 72°$ $180 - 72 = 108°$ $5 \times 108° = 540°$

Method 2

➤ Divide the polygon into triangles.

➤ Use the 'angle sum of a triangle' rule to work out the sum of all the angles.

➤ If it is a regular shape, then you can calculate the size of each angle.

$180 \times 3 = 540°$

 108°
$\frac{540}{5} = 108°$

1. One angle of an isosceles triangle is 62°. Find the other angles. (Note: there are two possible answers.)

2. Sketch an arrowhead (a special sort of kite), which has three of its angles 20°, 80° and 20°.

3. Calculate the external angle of a regular decagon.

4. If the internal angle of a regular polygon is 175°, how many sides has it got?

Similar and congruent shapes

Shapes are **similar** if they have corresponding angles equal and corresponding sides in the same ratio.
Shapes are **congruent** if they have corresponding angles and sides equal.

> Congruent shapes are the same shape **and** size.

Conditions for similarity

To show two shapes are similar, you must either:

➤ show that all the corresponding pairs of angles are equal

or

➤ show that all the corresponding pairs of sides are in the same ratio.

> This would mean that one shape is an enlargement of the other.

Area and volume in similar shapes

If two 3D shapes are similar, then one will be an enlargement of the other. If the linear scale factor of the enlargement is k, then the area scale factor will be k^2 and the volume scale factor will be k^3.

Conditions for congruency

To prove that two triangles are congruent, you must show that one of the following conditions is true.

> S – side
> A – angle
> R – right angle
> H – hypotenuse

Side–Side–Side (SSS)	Side, Angle, Side (SAS)	Angle, Corresponding Side, Angle (ASA)	Right angle, Hypotenuse, Side (RHS)
All three sides of one triangle are equal to the three sides in the other triangle.	Two sides and the included angle of one triangle are equal to the two sides and the included angle in the other triangle.	Two angles and a corresponding side of one triangle are equal to two angles and a corresponding side of the other triangle.	Both triangles have a right angle, an equal hypotenuse and another equal side.
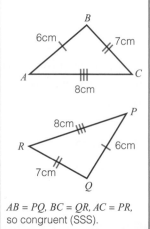	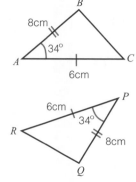	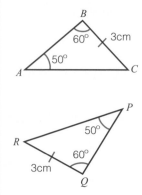	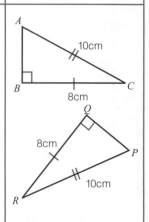
$AB = PQ$, $BC = QR$, $AC = PR$, so congruent (SSS).	$AB = PQ$, $AC = PR$, angle A = angle P, so congruent (SAS).	Angle A = angle P, angle B = angle Q, $BC = QR$, so congruent (ASA).	Angle B = angle Q, $AC = PR$, $BC = QR$, so congruent (RHS).

Proof

You can use congruence to prove angle facts in an isosceles triangle.
An isosceles triangle has two equal sides: $AB = BC$
BD bisects the angle ABC therefore angle ABD = angle CBD
The side BD is common to both triangles.
Therefore SAS: triangle ABD is congruent to triangle CBD
Therefore angle BAD = angle BCD: the base angles are equal.

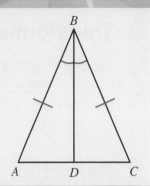

Create a colourful information poster to show the four conditions of congruency. Accurately draw a pair of triangles to illustrate each condition.

KEYWORDS

Similar ➤ Shapes with corresponding angles equal and corresponding sides in the same ratio.

Congruent ➤ Shapes with corresponding angles and sides equal.

1. Rectangle A has sides 3.4m and 5m. Rectangle B has sides 5.1m and 7.5m. Are the rectangles similar?

2. Are these pairs of triangles congruent? Give your reasons.

(a)

(b)
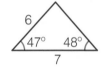

(c)
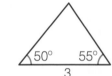

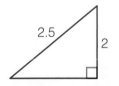

Transformations

All transformations transform an object to make an image. There are four different transformations that you need to know.

Reflection

Every reflection has a **mirror line**. The object is reflected so that the image is reversed and exactly the same distance from the line. ← A mirror image

Remember: To describe a reflection, you need to state the mirror line.

Translation

In a translation, every point slides the same distance across the plane. It is described with a **column vector**.

Remember: To describe a translation, you need to state the vector.

Rotation

Every point in the object is rotated through the same angle about a fixed point.

Remember: To describe a rotation, you need to state the centre of rotation, the angle and the direction.

Enlargement

Enlargement is the only transformation that can give an image which is a different size. The image and the object will always be similar. If the **scale factor** is p, the image of any point will be p times further from the centre of enlargement. ←

The image and the object are similar shapes.

Remember: To describe an enlargement, you need to state the centre of enlargement and the scale factor.

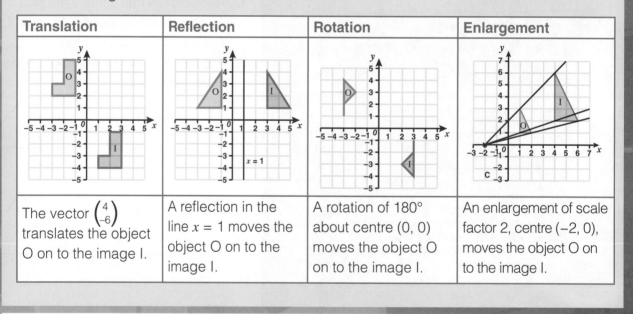

Translation	Reflection	Rotation	Enlargement
The vector $\binom{4}{-6}$ translates the object O on to the image I.	A reflection in the line $x = 1$ moves the object O on to the image I.	A rotation of 180° about centre (0, 0) moves the object O on to the image I.	An enlargement of scale factor 2, centre (−2, 0), moves the object O on to the image I.

Tricky enlargements

Enlargements can have fractional or negative scale factors:

➤ A scale factor less than 1 gives a smaller image that is closer to the centre of enlargement.

➤ A negative scale factor gives an image that is 'the other side' of the centre of enlargement.

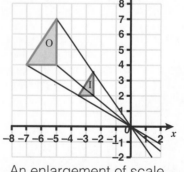

An enlargement of scale factor $\frac{1}{2}$, centre (0, 0)

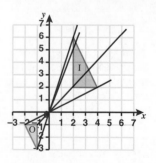

An enlargement of scale factor −2, centre (0, 0)

Module 28

Transformations

Combining different transformations

The order of the transformations matters. For example a rotation followed by a reflection **usually** gives a different image than doing the reflection first and then the rotation. You will be asked to do two transformations and then look for the **single** transformation that has the same effect.

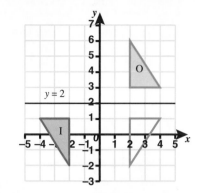

A reflection in the line $y = 2$ followed by a reflection in $x = 0$ is equivalent to a single rotation of 180° about (0, 2).

28

1. Describe fully each transformation from green to red.

(a)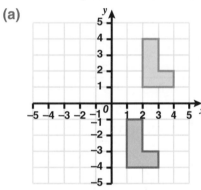

(b)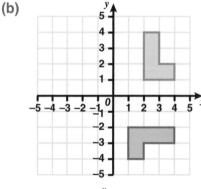

(c)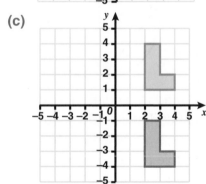

(d)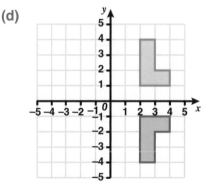

2. What is the **inverse** (opposite) of each of these transformations?
   (a) A reflection in line $y = 2$
   (b) A rotation of 90° clockwise about (2, 1)
   (c) A translation through $\begin{pmatrix} 7 \\ -2 \end{pmatrix}$
   (d) An enlargement of scale factor 4, centre (1, 1)
   (e) An enlargement of scale factor −4, centre (1, 1)
   (f) A rotation of 180° about (2, 10)

Use a desk light or similar bright light and a pencil to make a shadow on the table. Can you adjust the position of the object (the pencil) to make the image (the shadow) twice as big? Can you get an enlargement of scale factor 3?

Circle definitions

You need to know the correct names for the parts of a circle.

The parts of a circle are shown in these diagrams.

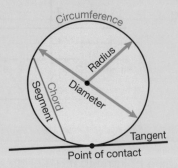

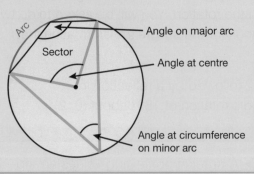

Create an information wall using sheets of A3 paper stuck together. Draw nine blank circles and illustrate the nine circle rules.

Circle theorems

You will need to know all these rules to solve circle geometry questions.

Two radii and a chord make an isosceles triangle. $OA = OB$ 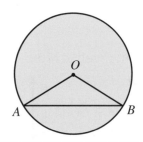	The angle between a tangent and a radius is always 90°. 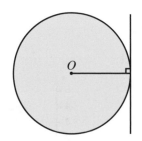	Two tangents from an external point to a circle are always the same length. $AB = AC$ 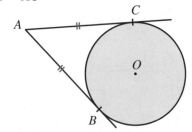
The **perpendicular** from the centre to a chord always bisects the chord. $AM = MB$ 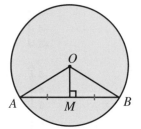	Angles in the same segment **subtended by an arc** are equal. angle a = angle b 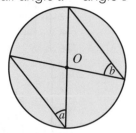	The angle subtended at the circumference by a semicircle equals 90°. angle c = 90° 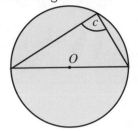
Angle subtended by an arc at the centre is twice the angle subtended at any point on the circumference. angle a = 2 × angle b 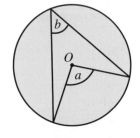	Opposite angles in a **cyclic quadrilateral** add up to 180°. angle a + angle c = 180° angle b + angle d = 180° 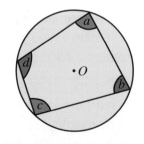	**Alternate segment theorem** angle ABC = angle CAQ and angle ACB = angle BAP 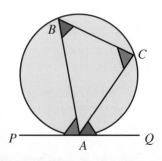

Proof

You might have to prove one of the angle theorems by writing out a set of logical steps.

1 The angle subtended at circumference by a semicircle equals 90°

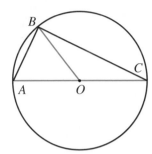

The three green lines are all radii so

$OA = OB = OC$

$\triangle OAB$ is isosceles

$\triangle OBC$ is isosceles

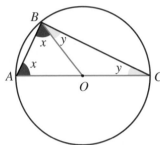

Each triangle contains a pair of equal angles.

Angles in $\triangle ABC$ sum to 180°

$\Rightarrow x + x + y + y = 180°$

$\Rightarrow x + y = 90°$

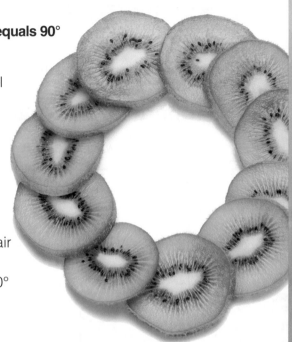

2 The alternate angle theorem

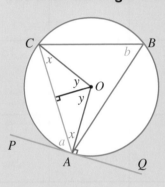

$\triangle OCA$ is isosceles so base angles x are equal.

Radius meets chord AC at 90° so two congruent triangles.

$\angle COA = 2y$

$x + y = 90°$ ← Angle sum of triangle = 180°

$\Rightarrow \angle CBA = b = \dfrac{2y}{2} = y$ ← Angle at centre = 2 × angle at circumference

$\angle OAP = 90°$ ← Radius and tangent are perpendicular

$\Rightarrow \angle CAP = a = 90 - x$

$\Rightarrow \angle CAP = 90 - (90 - y) = y$

$\Rightarrow \angle CAP = \angle CBA$

Theorem proved.

KEYWORDS

$\Rightarrow$ ➤ Symbol for 'implies' used in proofs to show how one step leads to the next.

Perpendicular ➤ Lines that meet at 90°.

Subtended by an arc ➤ Angle formed by two radii and an arc.

Cyclic quadrilateral ➤ A quadrilateral with all four vertices on the circumference of a circle.

---

1. Find the missing angles.

(a)

(b)

(c)

(d)

3D shapes

A **3D** shape can be described by its **faces**, **edges** and **vertices** (singular 'vertex').

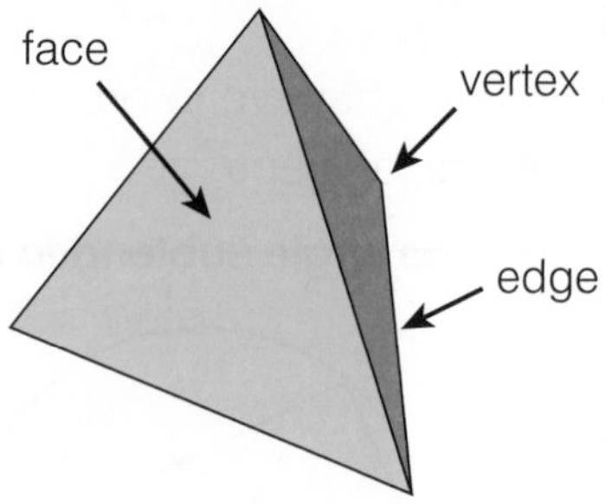

You may have to work out the area of a face. ← 2D measured in square units like cm²

You may have to work out the volume of the whole shape.

↑ 3D measured in cubic units like cm³

face → vertex → edge

Properties of 3D shapes

You need to know, or be able to work out, the properties of these 3D shapes.

	Shape	Name	Vertices	Faces	Edges
Prisms	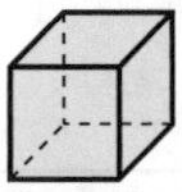	Cube	8	6	12
	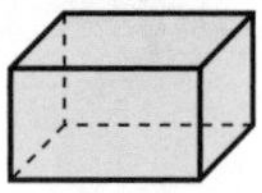	Cuboid	8	6	12
	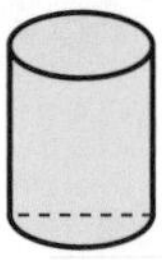	Cylinder	0	3	2
		Triangular prism	6	5	9
	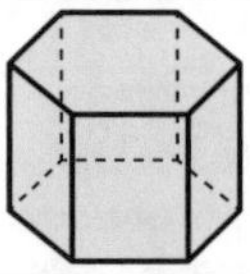	Hexagonal prism	12	8	18
Pyramids	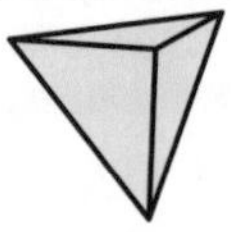	Tetrahedron	4	4	6
	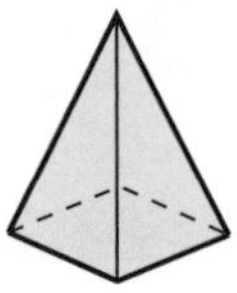	Square-based pyramid	5	5	8
	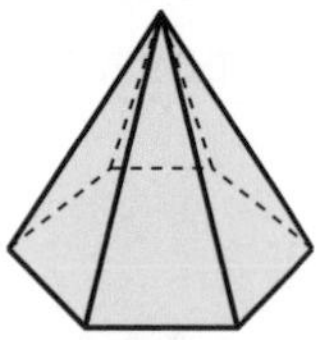	Hexagonal-based pyramid	7	7	12
Curved faces		Cone	1	2	1
		Frustum	0	3	2
	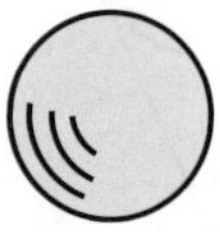	Sphere	0	1	0

3D ➤ Three dimensions. An object that has length, width and height.

Face ➤ A polygon that makes one surface of a 3D shape.

Edge ➤ Where two faces meet on a 3D shape.

Vertex ➤ The corner of a 2D or 3D shape.

Plan and elevation

Plans and elevations allow accurate 2D representation of 3D shapes.
They are widely used by architects and designers.
The plan is what you would see if you could hover above the shape – it is the outline.
The elevations are what you would see if you stood beside the shape – they are the outlines.

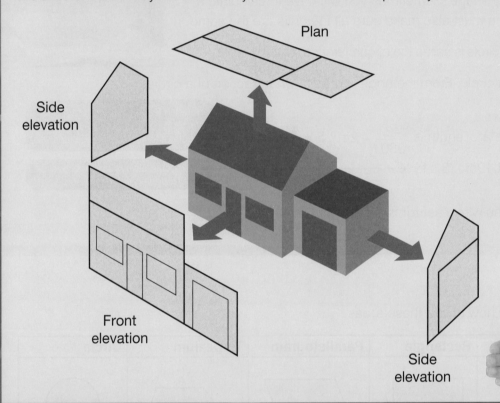

Plan

Side elevation

Front elevation

Side elevation

The Mobius strip – 'every rule has an exception'
Cut a long strip off a sheet of A4 (or larger) stiff paper or card. Twist one short edge by 180° and stick the two short edges together with tape. How many faces and edges has your new shape got? Is it 3D?

1. How many faces, edges and vertices do each of these 3D shapes have?
   (a) A hemisphere
   (b) A pentagonal prism
   (c) A pentagonal-based pyramid

2. Draw the plan and elevations of a hexagonal prism.

3. What shape has a plan and all elevations that are identical circles?

Perimeter, area and volume

Perimeter is a one-dimensional measure. It is a length measured in cm, m, km, inches, miles, etc.

Area is a two-dimensional measure. It is measured in cm², m², km², etc.

Volume is a three-dimensional measure. It is measured in cm³, m³, km³, etc.

Perimeter

The perimeter is 'all the way round the edge' so make sure you add up all the sides of any shape. Sometimes you will have to calculate a missing length. You must also make sure all the units are the same.

The perimeter of a circle is called the circumference, $C = 2\pi r$ or πd

An arc is part of a circle. From the angle work out what fraction of a circle.

The arc length is $\dfrac{48}{360} \times 2\pi \times 5 = \dfrac{4}{3}\pi\,\text{cm}$ ◄

> You can leave your answer like this 'in terms of π' or you can use your calculator to find a numerical answer.

$= 4.19\text{cm (3 s.f.)}$ ◄

> You should always state the level of rounding you have used.

The perimeter of the whole **sector** is $\dfrac{4}{3}\pi + 10\text{cm}$ or 14.2cm (3 s.f.)

Area

You need to know how to find these areas.

Triangle	Rectangle	Parallelogram	Trapezium	Circle
$A = \dfrac{1}{2}bh$	$A = bh$	$A = bh$	$A = \dfrac{1}{2}(a + b)h$	$A = \pi r^2$

h is the vertical height, at right angles to the base.

You must make sure all the units are the same and write the correct 'square unit' with your answer.

Find areas of **composite shapes** by chopping them up into any of the simple shapes above.

The area of the sector in the previous section is $\dfrac{48}{360} \times \pi \times 5^2 = \dfrac{10}{3}\pi\,\text{cm}^2$ or 10.5cm^2 (3 s.f.)

> The two area formulae below are given in your exam, so you will not have to learn them, but you must understand them and be able to use them correctly:
>
> **Curved surface area of a cone** $= \pi r l$ (where l is the slant height of the cone)
>
> **Surface area of a sphere** $= 4\pi r^2$

Volume

You need to know how to find these volumes.

Any prism	Any pyramid
Area of the cross-section × height (or length)	$\frac{1}{3}$ area of the cross-section × height
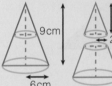	
Volume = 3 × 4 × 7 = 84cm³	Volume = $\frac{1}{3}$ × 3 × 4 × 7 = 28cm³

Find more complicated volumes by chopping them up into any of the simple volumes above. A frustum can be made by chopping the top off a cone.

The volume of this frustum is V.

> You can leave your answer like this 'in terms of π' or you can use your calculator to find a numerical answer.

$V = \frac{1}{3}\pi R^2 H - \frac{1}{3}\pi r^2 h$

$= \frac{1}{3}\pi[(6^2 \times 9) - (2^2 \times 3)]$

$= 104\pi\,\text{cm}^3$

$= 327\,\text{cm}^3$ (3 s.f.)

> You should always state the level of rounding you have used.

These two volume formulae are given in your exam, so you will not have to learn them, but you must understand them and be able to use them correctly:

Volume of a sphere $= \frac{4}{3}\pi r^3$

Volume of a cone $= \frac{1}{3}\pi r^2 h$

Make three larger copies of this net. Fold them to make three identical pyramids that will fit exactly together to make a cube.

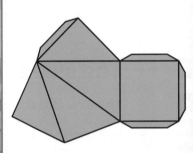

1. Find the perimeter and area of these shapes.

    (a)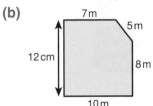

    (b)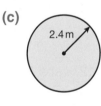

    (c)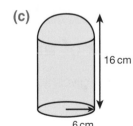

2. Find the volume of these shapes.

    (a)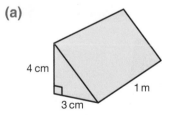

    (b)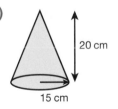

    (c)

3. Find the perimeter and the area of this major sector.

    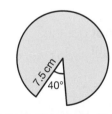

Pythagoras' theorem

Use Pythagoras' theorem for finding the missing side in a right-angled triangle when you have been given the other two lengths.

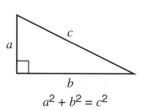

$$a^2 + b^2 = c^2$$

Basic trigonometry ratios

Basic trigonometry ratios only work for right-angled triangles. They are used to find angles from sides and sides from an angle and a side. Every angle has its own set of sin/cos/tan values.

$$\text{sine}\,\theta = \frac{\textbf{o}\text{pposite}}{\textbf{h}\text{ypotenuse}} \quad \left(s = \frac{o}{h}\right) \qquad \textbf{c}\text{osine}\,\theta = \frac{\textbf{a}\text{djacent}}{\textbf{h}\text{ypotenuse}} \quad \left(c = \frac{a}{h}\right) \qquad \textbf{t}\text{angent}\,\theta = \frac{\textbf{o}\text{pposite}}{\textbf{a}\text{djacent}} \quad \left(t = \frac{o}{a}\right)$$

> You could learn a mnemonic to help you remember them. Can you make up a better mnemonic than 'SOHCAHTOA'? Silly Old Henry Can…

The sine and cosine rules

The sine and cosine rules are harder to learn but they work on **any** triangle.

In a triangle labelled like this:

The sine rule: $\dfrac{a}{\sin A} = \dfrac{b}{\sin B} = \dfrac{c}{\sin C}$

or the other way up $\dfrac{\sin A}{a} = \dfrac{\sin B}{b} = \dfrac{\sin C}{c}$

The cosine rule: $a^2 = b^2 + c^2 - 2bc \cos A$

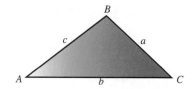

Using the sine and cosine rules in a triangle

Find angle P in this triangle.

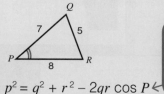

> Your triangle is labelled PQR and you need angle P so rewrite the equation carefully.

$p^2 = q^2 + r^2 - 2qr \cos P$

$5^2 = 8^2 + 7^2 - 2 \times 8 \times 7 \times \cos P$ ← Substitute in the values.

$\Rightarrow 25 = 113 - 112 \times \cos P$

$\Rightarrow \cos P = \dfrac{113 - 25}{112} \Rightarrow P = 38.2°$ (3 s.f.)

Find FH in this triangle.

Missing angle $180 - (86 + 42) = 52°$

$\dfrac{g}{\sin G} = \dfrac{f}{\sin F}$ ← Rewrite the sine rule equation carefully.

$\Rightarrow \dfrac{FH}{\sin 52°} = \dfrac{4.2}{\sin 86°}$

$\Rightarrow FH = \dfrac{4.2}{\sin 86°} \times \sin 52° = 3.3177…$

$FH = 3.32\text{m}$ (3 s.f.)

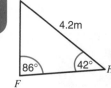

> Be careful using your calculator. You may have to key in sin(52) with both brackets.

The area of any triangle

This equation gives the area of any triangle:

Area $= \frac{1}{2}\, ab \sin C$

This is useful when you have **not** got the vertical height (so $\dfrac{\text{base} \times \text{height}}{2}$ won't work).

KEYWORDS

Hypotenuse ➤ The longest side in a right-angled triangle.

Opposite ➤ In a right-angled triangle, the side opposite the angle you are working with.

Adjacent ➤ In a right-angled triangle, the side next to the angle you are working with.

Using exact values for sin, cos and tan

Your calculator will give you decimal values of sin, cos and tan but you can also work out exact values for some special angles.

Angle	Triangle	sin	cos	tan
0°	As angle gets closer to 0° the opposite side approaches 0 and the adjacent side approaches 1.	$\dfrac{\text{opp}}{\text{hyp}}$ $= \dfrac{0}{1} = 0$	$\dfrac{\text{adj}}{\text{hyp}}$ $= \dfrac{1}{1} = 1$	$\dfrac{\text{opp}}{\text{adj}} = \dfrac{0}{1} = 0$
30°	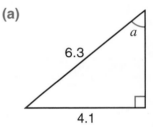 $1^2 + x^2 = 2^2 \quad x = \sqrt{3}$	$\dfrac{1}{2}$	$\dfrac{\sqrt{3}}{2}$	$\dfrac{1}{\sqrt{3}}$
60°		$\dfrac{\sqrt{3}}{2}$	$\dfrac{1}{2}$	$\sqrt{3}$
45°	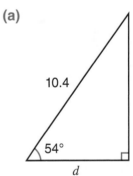 $1^2 + 1^2 = h^2 \quad h = \sqrt{2}$	$\dfrac{1}{\sqrt{2}}$	$\dfrac{1}{\sqrt{2}}$	1
90°	As angle gets closer to 90° the adjacent side approaches 0 and the opposite side approaches 1.	$\dfrac{\text{opp}}{\text{hyp}}$ $= \dfrac{1}{1} = 1$	$\dfrac{\text{adj}}{\text{hyp}}$ $= \dfrac{0}{1} = 0$	$\dfrac{\text{opp}}{\text{adj}} = \dfrac{1}{0}$ This is 'undefined' (your calculator will indicate an error if you ask it for tan 90°).

Draw any right-angled triangle accurately. Draw a square on each side like this. Cut out the two smaller squares. Can you cut them up to fit exactly inside the large square?

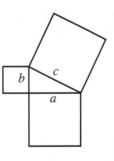

1. Is a triangle with sides 4.2cm, 7cm and 5.6cm right-angled? Explain your answer.

2. Calculate the missing angles.

   (a)
   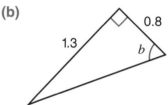

   (b)
   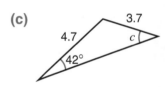

   (c)
   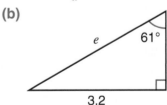

3. Calculate the missing lengths.

   (a)
   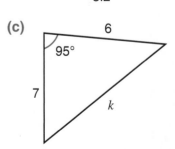

   (b)

   (c)

Vectors and translations

A vector can be written as a column vector, as a single bold letter plus an arrow, or with letters and an arrow on the top. It has both length and direction.

On this graph (2, 1) is the fixed point, A.

The column vector $\begin{pmatrix} 1 \\ 3 \end{pmatrix}$ is a **translation** of any point +1 in the x-direction and +3 in the y-direction. On this graph it is the movement from A to C and can also be written as **a** (with an arrow to show direction) or $\overrightarrow{AC}$.

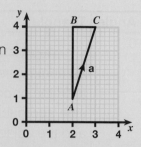

Working with vectors

You can add and subtract vectors. Think of this as one translation followed by another.

You can multiply any vector by a **scalar** which makes it larger or smaller.

a	b	a + b	a – b	2a	$\frac{1}{2}$ b	2a – b
$\begin{pmatrix} 1 \\ 2 \end{pmatrix}$	$\begin{pmatrix} 2 \\ -3 \end{pmatrix}$	$\begin{pmatrix} 1 \\ 2 \end{pmatrix} + \begin{pmatrix} 2 \\ -3 \end{pmatrix} = \begin{pmatrix} 3 \\ -1 \end{pmatrix}$	$\begin{pmatrix} 1 \\ 2 \end{pmatrix} - \begin{pmatrix} 2 \\ -3 \end{pmatrix} = \begin{pmatrix} -1 \\ 5 \end{pmatrix}$	$\begin{pmatrix} 2 \\ 4 \end{pmatrix}$	$\begin{pmatrix} 1 \\ -1.5 \end{pmatrix}$	$\begin{pmatrix} 2 \\ 4 \end{pmatrix} - \begin{pmatrix} 2 \\ -3 \end{pmatrix} = \begin{pmatrix} 0 \\ 7 \end{pmatrix}$

The **inverse** of a vector $\begin{pmatrix} a \\ b \end{pmatrix}$ is the vector $\begin{pmatrix} -a \\ -b \end{pmatrix}$

If you have to describe a vector 'in terms of' other vectors, just find your way between the points using the vectors you already have.

X and Y are the midpoints of AB and AC respectively.

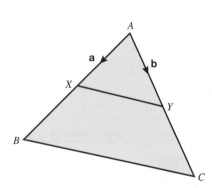

$\overrightarrow{XY} = -\mathbf{a} + \mathbf{b}$ or $\mathbf{b} - \mathbf{a}$ ← Vectors obey the same rules as numbers.

$\overrightarrow{YX} = -\mathbf{b} + \mathbf{a}$ or $\mathbf{a} - \mathbf{b}$

$\overrightarrow{AB} = 2\mathbf{a}$ $\overrightarrow{BA} = -2\mathbf{a}$

$\overrightarrow{AC} = 2\mathbf{b}$ $\overrightarrow{CA} = -2\mathbf{b}$

$\overrightarrow{BC} = -2\mathbf{a} + 2\mathbf{b}$

Module 33 Vectors

Vectors and proof

Vectors can be used to prove geometric facts.

In the diagram of a parallelogram
M and N are midpoints of the lines.
You can prove that CA is parallel to MN.

Find vectors $\overrightarrow{CA}$ and $\overrightarrow{MN}$ in terms of **a** and **c**
and then compare them.

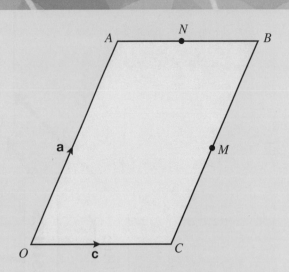

$\overrightarrow{CA} = -\mathbf{c} + \mathbf{a}$

$\overrightarrow{CB} = \overrightarrow{OA} = \mathbf{a}$

$\overrightarrow{AB} = \overrightarrow{OC} = \mathbf{c}$

$\overrightarrow{MN} = \dfrac{1}{2}\mathbf{a} - \dfrac{1}{2}\mathbf{c}$

Comparing: $\overrightarrow{CA} = 2 \times \overrightarrow{MN}$ so the vectors are parallel (but different lengths).

Use vectors to describe your movement around a room. You will need to decide which is the x-direction and which is the y-direction. For example, $\begin{pmatrix} 3 \\ 4 \end{pmatrix}$ could be three steps to the right and four steps forwards.

KEYWORDS

Translation ➤ A transformation where every point moves the same amount (a slide).

Scalar ➤ A number (with no units or dimensions).

Inverse ➤ The opposite, e.g. the inverse of +2 is −2 and the inverse of $\begin{pmatrix} 3 \\ -4 \end{pmatrix}$ is $\begin{pmatrix} -3 \\ 4 \end{pmatrix}$.

1. If $\mathbf{p} = \begin{pmatrix} 7 \\ -1 \end{pmatrix}$, $\mathbf{q} = \begin{pmatrix} 2 \\ 2 \end{pmatrix}$ and $\mathbf{r} = \begin{pmatrix} 0 \\ -1 \end{pmatrix}$ calculate:

   (a) $\mathbf{p} + \mathbf{r}$ (b) $\mathbf{p} + \mathbf{q} - \mathbf{r}$
   (c) $3\mathbf{p} + 2\mathbf{q}$ (d) $\mathbf{p} - 2\mathbf{r}$

2. $OABC$ is a quadrilateral, $\overrightarrow{OA} = \mathbf{a}$, $\overrightarrow{OB} = \mathbf{b}$ and $\overrightarrow{OC} = \mathbf{c}$. P, Q, R and S are the midpoints of OA, AB, BC and OC respectively.

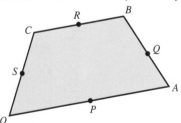

   Write $\overrightarrow{OP}$, $\overrightarrow{AB}$ and $\overrightarrow{CR}$, in terms of **a**, **b** and **c**.

33

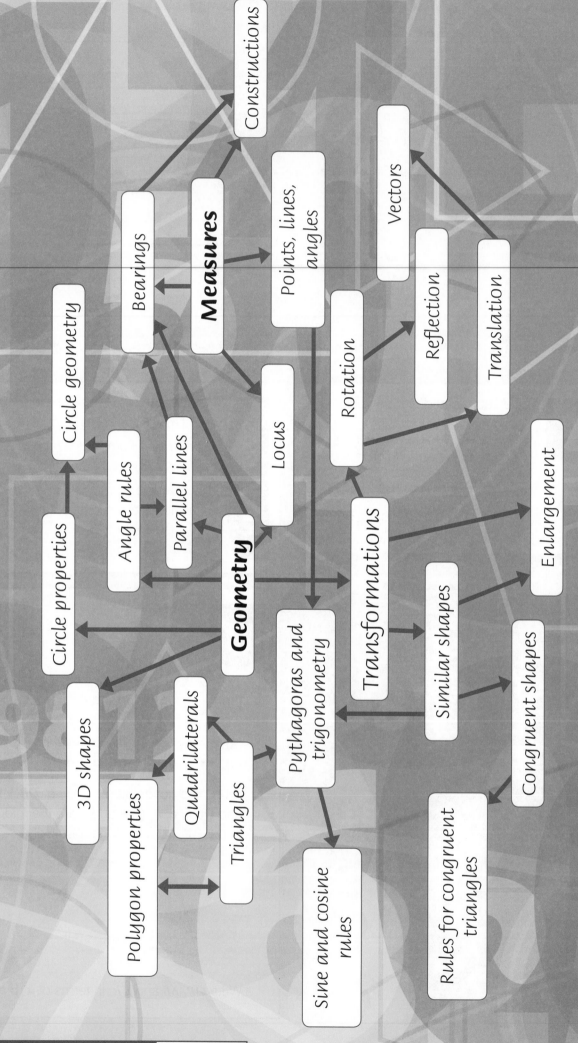

1. If the interior angle of a regular polygon is 156°, how many sides does it have? [2]

2. Find the angle HGD giving all your reasons. [3]

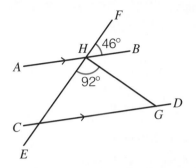

3. QS and RS are tangents to the circle with centre O.

   Calculate the angle POQ. [3]

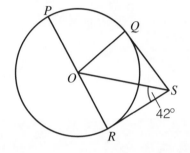

4. In triangle ABC the length AB = 10cm, angle BAC = 92° and angle ABC = 20°.
   In triangle RST the length SR = 10cm, angle RST = 20° and angle STR = 68°.

   Show that triangles ABC and RST are congruent, giving your reasons. [3]

5. Rotate shape A 90° clockwise about the point (0, 0) and label the image B. Reflect B in the line x = 0 and label the image C.

   What single transformation moves shape C on to shape A?

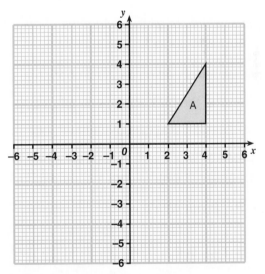

[4]

6. Calculate the area of triangle ABC if AB = 7cm, BC = 6cm and angle ACB = 64°. [4]

7. Find the angle NMQ. Give your answer correct to 3 significant figures. [3]

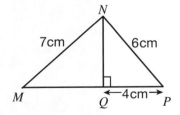

Theoretical probability

Probabilities can be written as fractions, decimals or percentages.
The probability of an event must be between 0 and 1 on the probability scale.

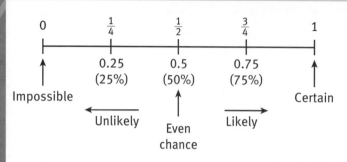

To find the probability of an event, use the formula:

$$P(\text{event}) = \frac{\textbf{Number of successful outcomes}}{\textbf{Total number of outcomes}}$$

P(drawing a king from a deck of cards) = $\frac{4}{52}$ ← 4 kings in a deck
← 52 cards in a deck

Mutually exclusive events

Mutually exclusive events cannot happen at the same time. For example, when you roll a dice, you can't get a 1 and a 6 at the same time.

The probabilities of all the outcomes of an event must add up to 1.

The probability of getting each number on a dice is $\frac{1}{6}$ and the probabilities all add up to 1 since it is certain that one of them will occur.

If the probability of event A happening is P, then the probability of event A **not** happening is 1 – P.

P(A) = P   P(not A) = 1 – P

If the probability of a white Christmas is 0.23, then the probability of not having a white Christmas is

1 – 0.23 = 0.77 ← The total probability of the two events must equal 1.

P(event) ➤ Probability of an event happening.

Mutually exclusive ➤ Events that can't happen at the same time.

Frequency ➤ The number of times something happens.

Biased ➤ Unfair, i.e. weighted on one side.

34

Relative frequency

It is not always possible to calculate a theoretical probability. In such cases estimating probability can be done by doing experiments or trials – this is called relative **frequency**.

This **biased** spinner is spun 20 times and then 100 times. Here are the results:

Score	1	2	3	4	Total trials
First test	2	3	1	14	20
Second test	11	14	7	68	100

$\leftarrow$ P(4) = $\frac{14}{20}$ = 0.7

$\leftarrow$ P(4) = $\frac{68}{100}$ = 0.68

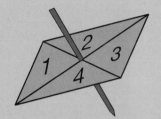

The more trials that are done, the more accurate the estimated probability will be. 0.68 is a better estimate than 0.7 for the probability of getting a 4 on this biased spinner.

Expected probability

If the probability of an event is known, then it is easy to predict the number of times that the event is expected to happen. Use the formula:

Expected number of occurrences = Probability of event × Number of trials

The probability that a biased coin will land on heads is 0.24. If the coin is tossed 600 times, how many heads will be expected to occur?

Expected number of heads = 0.24 × 600 = 144

1. A fair dice is rolled. Find:
   (a) P(even) **(b)** P(prime) **(c)** P(factor of 6)
2. The probability that a person will pass their driving test is 0.46
   (a) Find the probability that a person will not pass their driving test.
   (b) 500 people take their test this month. How many do you expect to pass?
3. Mary and Max both do an experiment to find out whether a drawing pin will land point up. Mary drops the pin 20 times and Max drops the pin 50 times. Here are the results:

	Point up	Total trials
Mary	13	20
Max	34	50

   (a) Who has the most reliable results and why?
   (b) What is the best estimate for the probability that the pin will land point up?

Roll a dice 30 times and record your results. Do the results show what you would have expected?
Do you believe your dice is biased or fair?

Roll the dice again 60 times and record your results. Are the results now closer to what you would expect?

Sample space

A **sample space** diagram is a list or a table which shows all the possible outcomes of two or more **combined** events.

Two dice are rolled and the scores are added together:

The probability of scoring 9 is $\frac{4}{36}$

Dice 1						
6	7	8	⑨	10	11	12
5	6	7	8	⑨	10	11
4	5	6	7	8	⑨	10
3	4	5	6	7	8	⑨
2	3	4	5	6	7	8
1	2	3	4	5	6	7
	1	**2**	**3**	**4**	**5**	**6**

Dice 2

KEYWORDS

Sample space ➤ Shows all the possible outcomes of an experiment.

Combined ➤ Two or more events occur at the same time.

Independent ➤ One event does not affect another.

Conditional ➤ One outcome depends upon another.

Independent events

Independent events can happen at the same time and the occurrence of one event does not affect the occurrence of the other. For example, if you roll a dice and flip a coin, you can get a head and a six and one event does not affect the other.

Learn the following rules:

Rule	What you say	What you do
AND rule	P(A and B)	P(A) × P(B) ← **AND means times.**
OR rule	P(A or B)	P(A) + P(B) ← **OR means add.**

Find the probability of getting a head on a coin and a six on a dice.

P(head AND six) = P(head) × P(six)

$$= \frac{1}{2} \times \frac{1}{6} = \frac{1}{12}$$

Tree diagrams

A tree diagram is a useful way of showing all of the outcomes of two or more combined events. They are also an excellent tool for calculating probabilities.

A coin is tossed and the spinner is spun.

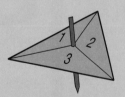

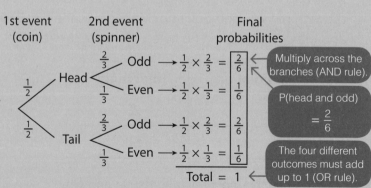

Find the probability of getting a head on the coin and an odd number on the spinner.

1st event (coin) — 2nd event (spinner) — Final probabilities

Head $\frac{1}{2}$

$\frac{2}{3}$ Odd → $\frac{1}{2} \times \frac{2}{3} = \frac{2}{6}$ ← Multiply across the branches (AND rule).

$\frac{1}{3}$ Even → $\frac{1}{2} \times \frac{1}{3} = \frac{1}{6}$

P(head and odd) = $\frac{2}{6}$

Tail $\frac{1}{2}$

$\frac{2}{3}$ Odd → $\frac{1}{2} \times \frac{2}{3} = \frac{2}{6}$

$\frac{1}{3}$ Even → $\frac{1}{2} \times \frac{1}{3} = \frac{1}{6}$

Total = 1 ← The four different outcomes must add up to 1 (OR rule).

Venn diagrams and set notation

Venn diagrams can also be used to represent and to calculate probabilities.

Some students took a music practical exam and a music theory exam. Everyone passed at least one exam. 46% passed the practical exam and 84% passed the theory exam. Find the probability that a student chosen at random passed the practical exam only.

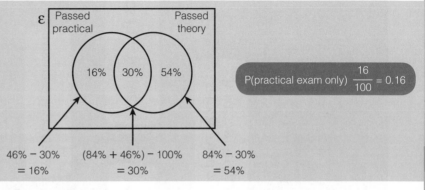

P(practical exam only) $\frac{16}{100} = 0.16$

46% − 30%
= 16%

(84% + 46%) − 100%
= 30%

84% − 30%
= 54%

This table shows some set notation which is represented by the Venn diagram below.

Set notation	Meaning	Set of numbers
$\mathcal{E} = \{x: 0 < x < 10\}$	Universal set = all items that could occur	{1, 2, 3, 4, 5, 6, 7, 8, 9}
A = $\{x: 0 < x < 10$ and x is odd}	Set A = odd numbers from 0 to 10	Set A = {1, 3, 5, 7 and 9}
B = $\{x: 0 < x < 10$ and x is prime}	Set B = prime numbers from 0 to 10	Set B = {2, 3, 5 and 7}

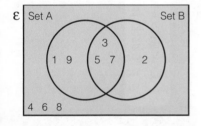

35

Conditional probability

Conditional probability is where the probability of an event depends upon the outcome of another event.

There are three chocolate raisins and four chocolate peanuts in a bag. Two chocolates are taken at random and eaten. Find the probability that at least one chocolate peanut is eaten.

1st chocolate   2nd chocolate

$\frac{2}{6}$ Raisin ⟶ $\frac{3}{7} \times \frac{2}{6} = \frac{6}{42}$

$\frac{4}{6}$ Peanut ⟶ $\frac{3}{7} \times \frac{4}{6} = \frac{12}{42}$

$\frac{3}{6}$ Raisin ⟶ $\frac{4}{7} \times \frac{3}{6} = \frac{12}{42}$

$\frac{3}{6}$ Peanut ⟶ $\frac{4}{7} \times \frac{3}{6} = \frac{12}{42}$

Only six chocolates are left after the 1st has been taken.

P(at least one peanut)

$= \frac{12}{42} + \frac{12}{42} + \frac{12}{42} = \frac{36}{42}$

Alternative method: 1 − P(two raisins) = $1 - \frac{6}{42} = \frac{36}{42}$

✓

1. Two dice are rolled and the scores are added together. Using the sample space diagram on page 80, find:
   (a) P(double 6) (b) P(not getting a double)
   (c) What is the most likely score to occur?

2. From the tree diagram on the left, find the probability of choosing two chocolates which are the same type.

3. There are 2 black and 3 white counters in a bag. One counter is taken at random and not replaced. A 2nd counter is then taken. Find the probability that at least one of the counters taken is black.

Flip a coin three times. Repeat the experiment 40 times. Record the number of times that you get three heads in a row.

Draw a tree diagram for three coins and compare your theoretical results with your experimental results. How close are they?

Probability

Theoretical
- Probability scale
- Fractions, decimals and percentages
- Equally likely outcomes
  - Random
  - Fair
    - Bias
- Mutually exclusive
  - Add up to 1
  - $P(A) = 1 - P(\text{not } A)$
- Combined events
  - Independent events
    - AND/OR rule
  - Conditional probability

Experimental
- Estimated
- Sample size
- Interpret
- Relative frequency

Representing
- Venn diagrams
- Tree diagrams
- Sample space
- Two-way tables
- Conditional probability

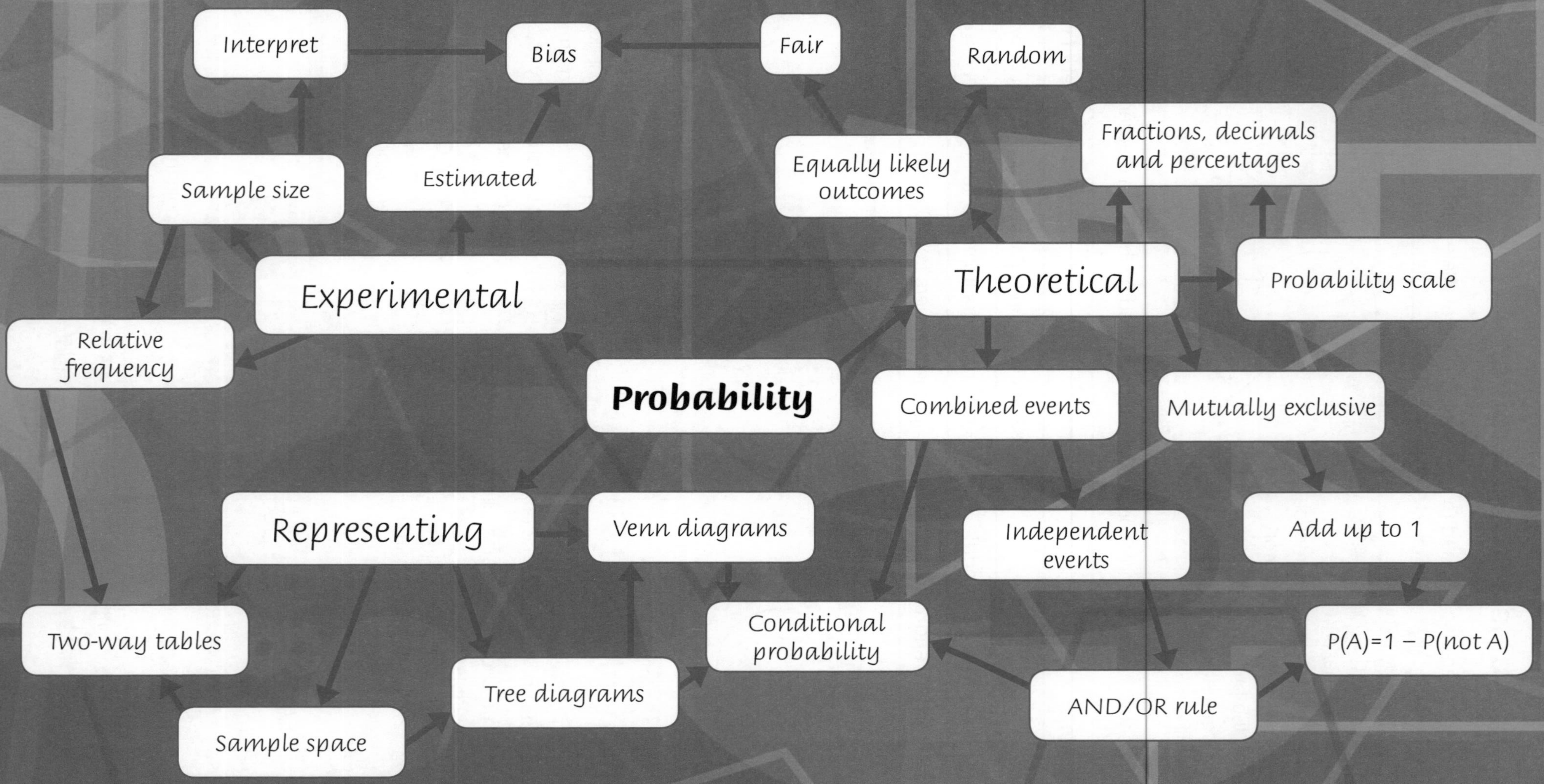

1. A letter is chosen at random from the word PROBABILITY. Find the probability that the letter will be: Ⓕ

   (a) A letter B **[1]**   **(b)** A vowel **[1]**

   (c) Not an I **[1]**   **(d)** A letter Q **[1]**

2. The probability that a biased dice will land on the numbers 1 to 5 is given in the table.

Number	1	2	3	4	5	6
Probability	0.1	0.2	0.25	0.1	0.15	

   (a) Find the probability that the dice will land on 6.   **[2]**

   (b) The dice is rolled 60 times. How many times do you expect it to land on 5?   **[2]**

3. Two fair coins are flipped. Find the probability of getting: Ⓕ

   (a) Two heads   **[2]**

   (b) At least one tail   **[2]**

4. Mark and Simon play snooker and then table tennis.
   The probability that Mark will win at snooker is 0.3
   The probability that Mark will win at table tennis is 0.6

   (a) Copy and complete the tree diagram.   **[2]**

   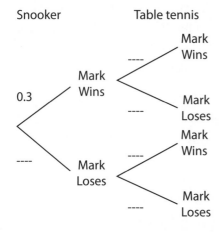

   Snooker   Table tennis

   0.3 Mark Wins
   ---- Mark Wins
   ---- Mark Loses
   ---- Mark Loses
   ---- Mark Wins
   ---- Mark Loses

   (b) Find the probability that Mark will win both games.   **[2]**

   (c) Find the probability that Mark will win at least one game.   **[2]**

5. 50 students were asked to try chocolate A and chocolate B. Here are the results: Ⓕ

   45 said they liked chocolate A   46 said they liked chocolate B   42 said they liked both

   (a) Draw a Venn diagram to show the information.   **[2]**

   (b) Find the probability that a student chosen at random did not like either.   **[2]**

   (c) Given that a student did like chocolate A, find the probability that they did
   not like chocolate B.   **[2]**

6. If it is a clear night, then the probability of frost in the morning is $\frac{7}{10}$

   If it is a cloudy night, then the probability of frost in the morning is $\frac{3}{10}$

   The probability that it is clear tonight is $\frac{3}{5}$

   Find the probability that there is frost in the morning. Ⓕ   **[4]**

Types of data

You need to know the difference between qualitative and quantitative data and between **discrete** and **continuous data**.

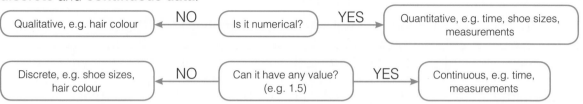

| Qualitative, e.g. hair colour | ← NO | Is it numerical? | YES → | Quantitative, e.g. time, shoe sizes, measurements |

| Discrete, e.g. shoe sizes, hair colour | ← NO | Can it have any value? (e.g. 1.5) | YES → | Continuous, e.g. time, measurements |

Averages

There are three types of average – **mean**, **median** and **mode**. These are used to help compare different populations.

36

Ungrouped data

The table below shows the goals scored by a school football team in 70 games.

Goals scored	Frequency	Total goals	
0	22	0 × 22 =	0
1	21	1 × 21 =	21
2	13	2 × 13 =	26
3	10	3 × 10 =	30
4	2	4 × 2 =	8
5	2	5 × 2 =	10
	70		95

↑ Total number of games

↑ Total number of goals scored in all the 70 games

$\text{Mean} = \dfrac{\text{Total goals scored}}{\text{How many games}}$

$= 95 \div 70 = 1.4$ goals per game

Median interval: The group with the middle value. So out of 70 games, halfway is between the 35th and 36th value. Both of these will be in the second group. So the median value is 1.

Mode: Most common number of goals scored is 0 (the group with the highest frequency). So the mode is 0.

Grouped data

The table below shows the times taken by 200 runners in a race.

Time (mins)	Frequency	Midpoint	Frequency × Midpoint
$10 \leqslant t < 15$	1	12.5	12.5
$15 \leqslant t < 25$	8	20	160
$25 \leqslant t < 35$	33	30	990
$35 \leqslant t < 45$	56	40	2240
$45 \leqslant t < 55$	48	50	2400
$55 \leqslant t < 65$	35	60	2100
$65 \leqslant t < 75$	19	70	1330
	200		9232.5

↑ Total number of runners

↑ This is the assumed running time for every person in that group.

↑ Total time taken by all the runners

Estimated mean

$= \dfrac{\text{Total time for all runners}}{\text{How many runners}}$

$= 9232.5 \div 200 = 46.2$ minutes

This is only an estimated mean as it uses the midpoint for each group and not the actual values of the times for each person.

Median interval: Between 100th and 101st runner – both these values will be in the fifth group. So the median will be in the group $45 \leqslant t < 55$.

Modal group: $35 \leqslant t < 45$ (the group with the highest frequency)

Data and Averages

Module 36

Sampling

To find out about a **population** or to test a **hypothesis**, a sample is taken and analysed.

How has it been collected? Does anything make it biased? · What makes a good sample? · Sample size 3 to 10% of the population

There are different methods for picking a sample (e.g. systematic – pick every tenth one or do it randomly by using a random number generator) and there are different sample types (e.g. stratified – the sample has the same proportion of each group as per the population).

For example, a supermarket chain wants to find out how much people spend on groceries in a week. It asks the first 30 people who shop at its Watford branch in December. This sample would be **biased** as it is limited to one store and so is not a **representative sample**. Also, 30 people is not a large enough sample.

Conduct your own survey on how many text messages people send in a week. Do this for both boys and girls separately. Use a table like this to record your data.

Number of texts	Boys			Girls		
	Freq.	Midpt.	F×M	Freq.	Midpt.	F×M
$0 \leqslant t < 20$						
$20 \leqslant t < 40$						
$40 \leqslant t < 60$						
$60 \leqslant t < 80$						
$80 \leqslant t < 100$						
$100 \leqslant t < 150$						
$150 \leqslant t < 200$						

Create a short video clip on how you calculate the mean for the boys and the mean for the girls.

Two-way tables

Two-way tables show two different variables. Examples include train and bus timetables.

Watford	07.35	08.05	08.35	09.05	10.05
Birmingham	09.25	09.55	10.25	10.55	11.55
Sheffield	10.23	10.53	11.23	11.53	12.23
Leeds	11.36	12.06	12.36	13.06	14.06
Sunderland	12.42	13.12	13.42	14.12	15.12
Newcastle	13.48	14.18	14.48	15.18	16.18

The columns represent the different trains. For example, there is a train leaving Watford at 10.05am and arriving in Leeds at 2.06pm.

1. When working with grouped data, why can you only find an estimate of the mean?
2. Adam has a hypothesis that girls spend more time surfing the Internet than boys. He takes a sample of 10 girls and 10 boys in his class. Is this a good sample? Why?
3. Look at the train timetable (left). Ruth needs to be in Newcastle for an interview at 3.00pm. It will take her 25 minutes to get to the interview from the train station once she arrives. She lives in Sheffield. What is the latest train she can catch and how long will it take?

KEYWORDS

Discrete data ➤ Can only have set values, e.g. hair colour, shoe size, trouser size.

Continuous data ➤ Can take any value, e.g. things you measure (time, length, weight).

Mean ➤ Total of values divided by the number of values.

Median ➤ The middle value when the data is ordered.

Mode ➤ The most common value.

Sampling ➤ A way of collecting data from a population.

Population ➤ The data set, e.g. people or objects.

Hypothesis ➤ An idea you believe to be true.

Bar charts, pictograms and line graphs

Bar charts, pictograms and vertical line graphs are best for discrete and qualitative data. As well as being able to draw them, you need to be able to read them and pick an appropriate diagram for the situation.

Pictograms use pictures to represent data and have a key to show how the picture relates to the frequency of the data.

The number of complaints at a call centre was recorded over a period of two weeks. Bar charts and line graphs can be used to compare this data.

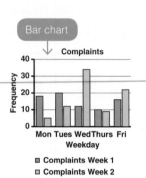

Bar chart

Complaints

□ Complaints Week 1
□ Complaints Week 2

	Complaints Week 1	Complaints Week 2
Mon	18	5
Tues	20	12
Wed	12	34
Thurs	10	9
Fri	16	22

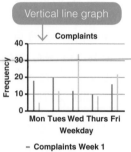

Vertical line graph

Complaints

– Complaints Week 1
– Complaints Week 2

Pie charts

Pie charts work best for discrete and qualitative data. Here are the results of a survey of 60 students from Year 7 on how many text messages they send in a week.

This is the number of people in each category.

Number of texts	Frequency	Angle
$0 \leqslant t < 50$	18	$360° \div 60 \times 18 = 108°$
$50 \leqslant t < 100$	20	$360° \div 60 \times 20 = 120°$
$100 \leqslant t < 200$	12	$360° \div 60 \times 12 = 72°$
$200 \leqslant t < 300$	10	$360° \div 60 \times 10 = 60°$

Survey of Year 7 students – text messages sent per week

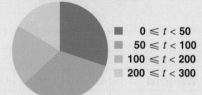

■ $0 \leqslant t < 50$
■ $50 \leqslant t < 100$
■ $100 \leqslant t < 200$
■ $200 \leqslant t < 300$

Remember that there are 360° in a circle. So divide 360 by the total frequency (60 students), then multiply by the frequency for each category.

Line graphs for time series data

Some types of data that are measured against time can be represented on a time series graph. Examples include prices, Retail Price Index, share prices, sales, etc. You will need to be able to plot and interpret these.

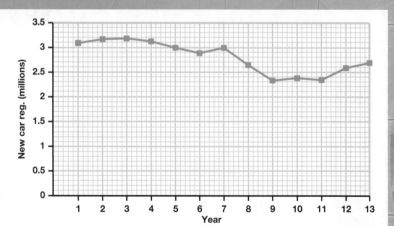

Statistical Diagrams

Module 37

Scatter graphs

Scatter graphs are used to test for a relationship between two variables. They will show positive **correlation**, negative correlation or no correlation.

Positive correlation

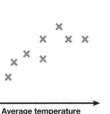

Negative correlation

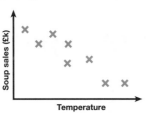

No correlation

A correlation can be weak or strong, depending on how closely the points follow the line of best fit. You can use a line of best fit to make predictions but only **within the range of the data**. It would be unreliable to extend the line of best fit and make predictions beyond the range of the data because the relationship may not hold.

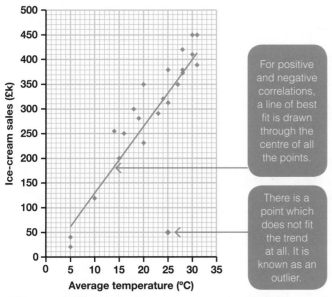

For positive and negative correlations, a line of best fit is drawn through the centre of all the points.

There is a point which does not fit the trend at all. It is known as an outlier.

Create your own pie chart on an A4 sheet of paper to show the number of pages in 40 books in your home.

Histograms

Histograms work best for grouped data and are similar to bar charts except:
➤ There are no spaces between the bars.
➤ The y-axis is **frequency density** rather than frequency.

Here are the times for 200 people to run a race.

Running time (mins)	Frequency	Frequency density (Frequency ÷ Class width)
$10 \leqslant t < 15$	10	$10 \div 5 = 2$
$15 \leqslant t < 20$	0	$0 \div 5 = 0$
$20 \leqslant t < 25$	35	$35 \div 5 = 7$
$25 \leqslant t < 35$	64	$64 \div 10 = 6.4$
$35 \leqslant t < 45$	40	$40 \div 10 = 4$
$45 \leqslant t < 55$	32	$32 \div 10 = 3.2$
$55 \leqslant t < 70$	19	$19 \div 15 = 1.3$
	200	

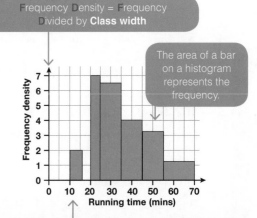

Frequency Density = Frequency Divided by **Class width**

The area of a bar on a histogram represents the frequency.

A continuous scale (i.e. equally spaced out and no gaps) and labelled with what the data is about (e.g. height, weight, time, etc.)

KEYWORDS

Correlation ➤ The relationship between two variables.

Frequency density ➤ The y-axis on a histogram, calculated by frequency divided by class width.

Class width ➤ How big the category is, e.g. in a frequency table the category $20 \leqslant t \leqslant 30$ would have a class width of 10.

1. Look at the time series graph on page 86.
   (a) What could have caused the dip in years 7–9 in new car registrations?
   (b) How many new car registrations would you predict for year 14?

Cumulative frequency graphs

Cumulative frequency graphs are used to find the median and **interquartile range** (IQR). The x-axis is a continuous scale and the y-axis has the **cumulative frequency**.

In a different race to that on page 87, the times of another 200 runners were recorded in this table.

It is the upper boundary for each category that is the x-value when plotting the points.

Time (mins)	Frequency	Cumulative frequency
$10 \leqslant t < 15$	1	1
$15 \leqslant t < 25$	8	9
$25 \leqslant t < 35$	33	42
$35 \leqslant t < 45$	56	98
$45 \leqslant t < 55$	48	146
$55 \leqslant t < 65$	35	181
$65 \leqslant t < 75$	19	200
	200	

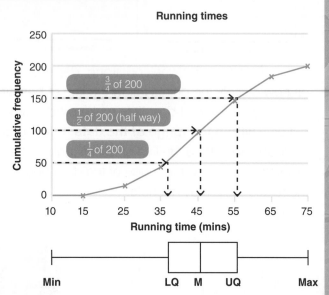

Box plots

Box plots show key points within the data, such as the median and how spread out the data is.

Using the example above for running times, assuming from the table that the quickest time is 10 minutes and the longest time is 75 minutes, you can construct a box plot as shown.

Quartiles and interquartile range

As well as the averages, the IQR is used to compare distributions. The IQR represents the middle 50% of a population.

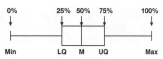

The interquartile range is calculated by upper quartile (UQ) – lower quartile (LQ).

A cumulative frequency diagram and a histogram can be used to find the median and interquartile range.

In a survey, the actual weights of 12 100g chocolate bars were recorded in grams: 97, 94, 102, 95, 97, 106, 105, 101, 99, 97, 106, 104.

Can you find the mean, median, mode and the IQR?

First put the values in order:

94, 95, 97, 97, 97, 99, 101, 102, 104, 105, 106, 106

Median is between 6th and 7th values = 100g

Mode: Most common value is 97g

Mean: (Total weight = 1203) ÷ 12 = 100.25g

IQR: 94, 95, 97, 97, 97, 99, 101, 102, 104, 105, 106, 106

Twelve values so:

UQ between 9th and 10th values so 104.5g

LQ between 3rd and 4th values so 97g

Hence IQR = 104.5 – 97 = 7.5g

Module 38

Comparing Distributions

Describing populations

To compare different populations you can use the averages and the ranges. The results of a survey on salaries have been plotted:

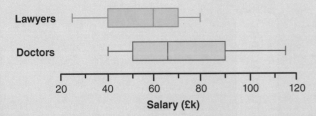

Lawyers

Doctors

How would you describe each distribution and what are the differences? Who earns more?

Lawyers: the median is £60k, IQR = 70 – 40 = £30k and looking at the box plot the distribution is skewed to the left.

Doctors: the median is £65k, IQR = 90 – 50 = £40k and looking at the box plot the distribution is skewed to the right.

The box plots suggest that doctors earn more because their median is higher, but their IQR is higher meaning that there is a greater variation in salaries for doctors than there is for lawyers.

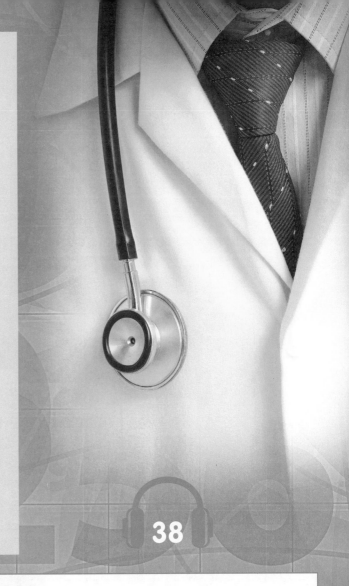

38

Create flashcards to make sure you know the meaning of all the keywords listed in Modules 36 to 38. Write the keyword on one side of the flashcard and the definition on the other. Write definitions using your own words but make sure they are accurate by checking against this book.

1. Consider the runners' data on page 88.
   (a) Calculate the frequency densities.
   (b) Use the cumulative frequency graph to estimate how many runners took more than 50 minutes.
2. (a) A sample of 400 people was asked about how much they spent a week on food. The lowest amount was £26 and the highest amount was £90. 25% said that they spent less than £55. The interquartile range was £22. The median amount spent was £66.
   Show this information on a box plot. (Don't forget you will need to draw a scale first.)
   (b) In a different survey, the median amount spent was £55 and the interquartile range was £35. Write two comments about the differences between the two surveys.

Statistics

Statistics

- Data
  - Discrete
  - Continuous
  - Two-way tables
  - Ungrouped
    - Pie charts
    - Bar charts
    - Vertical line graphs
  - Averages
    - Outliers
    - Mode
    - Median
    - Mean
    - Grouped (estimated mean)

- Compare distributions

- Statistical diagrams
  - Time series
  - Scatter graphs
    - Outlier
    - Line of best fit
    - Correlation

- IQR
  Median
  - Histograms
    Cumulative frequency diagrams
    Box plots

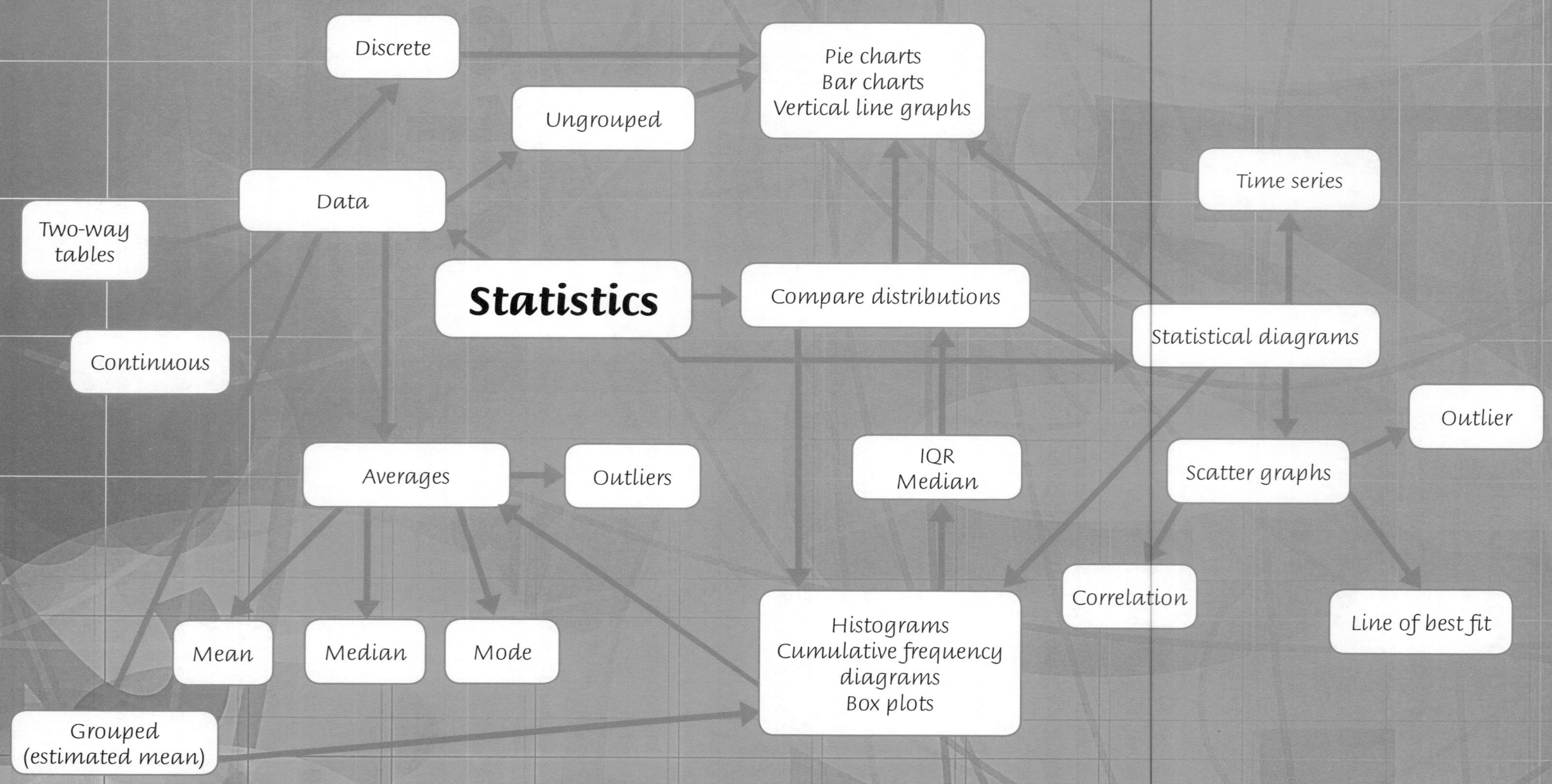

1. Look at this data showing the weights of 200 people.

Weight (kg)	$10 \leqslant w < 15$	$15 \leqslant w < 25$	$25 \leqslant w < 35$	$35 \leqslant w < 45$	$45 \leqslant w < 55$	$55 \leqslant w < 65$	$65 \leqslant w < 75$	
Frequency	36	24	13	31	40	33	23	200

    (a) Find an estimate of the mean.   **[4]**

    (b) Find the median interval.   **[1]**

    (c) Find the modal group.   **[1]**

2. Draw a histogram for the data in question 1.   **[3]**

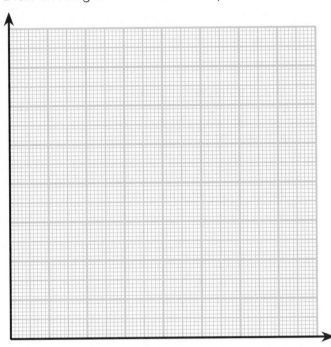

3. (a) Draw a cumulative frequency curve for the data in question 1.   **[3]**

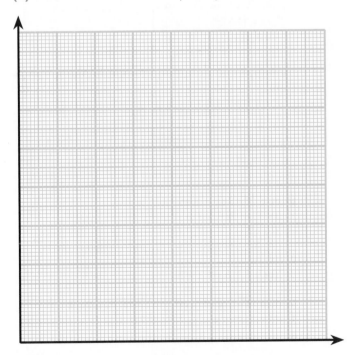

    (b) Estimate how many people weighed over 60 kg.   **[2]**

4. Five positive integers have a median of 5, a mean of 6 and a mode of 3.
   What is the greatest possible range for the numbers?   **[3]**

Module 1 – Place Value and Ordering

1. (a) +3   (b) –8   (c) –5
2. (a) 48   (b) –8   (c) –60
3. (a) False   (b) True   (c) True
4. (a) 2.451   (b) 245 100   (c) 430

Module 2 – Factors, Multiples and Primes

1. (a) Factors of 12: 1, 2, 3, 4, 6, 12   Factors of 15: 1, 3, 5, 15
   (b) 3
2. (a) 7, 14, 21, 28, 35, 42, 49, 56   5, 10, 15, 20, 25, 30, 35, 40
   (b) 35
3. (a) $28 = 2 \times 2 \times 7$ $98 = 2 \times 7 \times 7$
   (b) $28 = 2^2 \times 7$ $84 = 2 \times 7^2$
   (c) HCF = 14   LCM = 196

Module 3 – Operations

1. (a) 18   (b) 24   (c) 38   (d) 2
2. (a) $\dfrac{5}{2}$ (b) 7   (c) $\dfrac{1}{5}$ (d) 4
3. (a) $\dfrac{1}{20}$ (b) $5\dfrac{5}{8}$ (c) $\dfrac{20}{21}$ (d) $4\dfrac{2}{3}$

Module 4 – Powers and Roots

1. (a) (i) 3^9 (ii) 4^7 (iii) 6^7
   (b) (i) 1   (ii) $\dfrac{1}{125}$ (iii) 4
2. (a) (i) 6×10^3 (ii) 2.3×10^3 (iii) 6.78×10^{-3}
   (iv) 1.5×10^{-1}
   (b) (i) 1.8×10^{10} (ii) 2×10^6 (iii) 4.71×10^6
3. (a) $5\sqrt{2}$ (b) $4\sqrt{5} - 5$ (c) $5 - \sqrt{7}$

Module 5 – Fractions, Decimals and Percentages

1. (a) 0.6   (b) 0.875   (c) $0.\dot{6}$
2. (a) $\dfrac{9}{25}$ (b) $\dfrac{31}{125}$ (c) $\dfrac{3}{11}$
3. (a) 30.87   (b) 30   (c) 14.4

Module 6 – Approximations

1. (a) 2.39   (b) 4.610   (c) 44 000   (d) 0.004 03
2. 750
3. (a) LB = 22.5, UB = 23.5   (b) LB = 4.65, UB = 4.75
   (c) LB = 5.225, UB = 5.235
4. (a) 411.25m²   (b) 0.48 (to 2 d.p.)

Module 7 – Answers Using Technology

1. (a) 12.167   (b) 4096   (c) 8
2. $3.1\dot{6}$
3. (a) 9.88×10^3 (b) 7.46×10^{14}

Module 8 – Algebraic Notation

1. (a) $12ab + 2a^2b$
   (b) $9x - 9$
   (c) $2ab + 2ac + 2bc$
2. (a) g^8
   (b) p^5
   (c) $14p^2q^{-2}$
3. (a) 36
   (b) $\dfrac{1}{16}$
   (c) $\dfrac{27}{8}$

Module 9 – Algebraic Expressions

1. (a) $10x - 15x^2$
   (b) $6x^2 + x - 35$
2. (a) $(x + 7)(x - 8)$
   (b) $(2p + 5q)(2p - 5q)$

3. (a) $\dfrac{x - 4}{x + 6}$
   (b) $\dfrac{4x + 10}{(x + 2)(x + 4)}$

Module 10 – Algebraic Formulae

1. (a) $p = \dfrac{(5 - a)^2}{a + 2}$
   (b) $p = \dfrac{2 - 3q}{2q - 3}$
2. (a) $p = 84$
   (b) $p = -\dfrac{3}{10}$
3. (a) $\dfrac{80}{s + 10}$
   (b) 6m/s

Module 11 – Algebraic Equations

1. $x = -\dfrac{4}{3}, x = 5$
2. $x = 6 + \sqrt{18}, x = 6 - \sqrt{18}$
3. $x = -\dfrac{7}{3}, y = \dfrac{22}{3}$ and $x = 2, y = 3$

Module 12 – Algebraic Inequalities

1. $x < \dfrac{8}{3}$

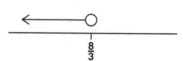

2. $(-10, 8)$

3.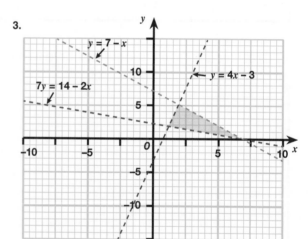

Module 13 – Sequences

1. (a) $U_n = 9n - 5$
   (b) 895
   (c) No, since $9n - 5 = 779$ has no integer solution.
2. (a) 45
   (b) $a = 1, b = 1, c = 3$

Module 14 – Coordinates and Linear Functions

1. $4y = 16x - 4$ and $4y = 8 - x$
2. $y = -\dfrac{x}{4} + 2$
3. $9x - 4y - 6 = 0$ or $y = \dfrac{9}{4}x - \dfrac{3}{2}$

Answers

Module 15 – Quadratic Functions

1. **(a)** $(-5, 0), (8, 0), (0, -40)$

    (b) $x = \dfrac{3}{2}$

    (c) $\left(\dfrac{3}{2}, -\dfrac{169}{4}\right)$

2. **(a)** $(-5, -65)$ **(b)** $x = -5$

Module 16 – Other Functions

1.

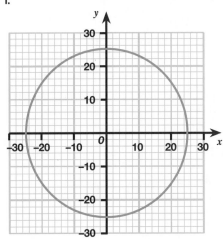

2. $fg(x) = 2x^2 + 8$, $gf(x) = 4x^2 + 24x + 37$, $f^{-1}(x) = \dfrac{x - 6}{2}$, $ff(x) = 4x + 18$, $gg(x) = x^4 + 2x^2 + 2$

Module 17 – Problems and Graphs

1.

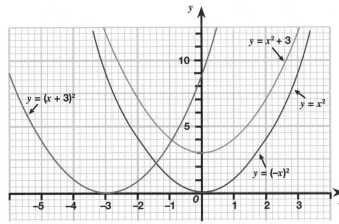

$-7x + 24y = 625$

2. The gradients at any point are equal to the x-coordinate at that point.

Module 18 – Converting Measurements

1. 17.8m/s (3 s.f.)   **2.** In UK €80 = £64 or £48 = €60
3. 84cm²   **4.** 2 500 000cm³

Module 19 – Scales, Diagrams and Maps

1. **(a)** 10km   **(b)** 25km   **(c)** 3.75km
    (d) 14cm   **(e)** 16cm   **(f)** 2cm
2. Accurately drawn diagrams to ±1°.
    (a) 265°   **(b)** 010°   **(c)** 120°

Module 20 – Comparing Quantities

1. **(a)** £8.40   **(b)** 392   **(c)** £19 200
2. **(a)** £19.04   **(b)** 579.12   **(c)** 8.1% (2 s.f.)
3. £384

Module 21 – Ratio

1. £16 000, $\dfrac{3}{7}$ **2.** 5.4m for £4.43 (82 p/m and 86 p/m)   **3.** 60

Module 22 – Proportion

1. £11.55   **2.** $\dfrac{14}{3}$ **3.** 1.25 (3 s.f.)

Module 23 – Rates of Change

1. **(a)** 1   **(b)** 0.5   **(c)** $-\dfrac{3}{7}$

2. **(a)** 2   **(b)** 5   **(c)** –5

Module 24 – Constructions

1. **(a)** Two lines parallel to and 3cm away from the drawn line (above and below), joined with semicircles radius 3cm at each end
    (b) Accurate perpendicular bisector of line ST
    (c) Accurate angle bisector
2. Accurate 60° angle with visible construction lines and angle bisector

Module 25 – Angles

1. $a = 52°$ (opposite angles), $b = 52°$ (corresponding/alternate angles), $c = 112°$ (alternate to 60° + 52°)
2. 109°   **3.** $x = 76°$, $y = 130°$, $z = 130°$

Module 26 – Properties of 2D Shapes

1. Either 62°, 56° or 59°, 59°   **2.** Missing angle 240°
3. 36°   **4.** 72

Module 27 – Congruence and Similarity

1. Yes (ratios both 1.5)
2. **(a)** Yes SAS or ASA   **(b)** Yes ASA   **(c)** Yes RHS

Module 28 – Transformations

1. **(a)** Translation $\begin{pmatrix} -1 \\ -5 \end{pmatrix}$ **(b)** A rotation of 90° clockwise about (0, 0)

    (c) Translation $\begin{pmatrix} 0 \\ -5 \end{pmatrix}$ **(d)** A reflection in line $y = 0$ (or the x-axis)

2. **(a)** A reflection in line $y = 2$
    (b) A rotation of 270° clockwise or 90° anti-clockwise about (2, 1)

    (c) A translation through $\begin{pmatrix} -7 \\ 2 \end{pmatrix}$

    (d) An enlargement of scale factor 0.25, centre (1, 1)
    (e) An enlargement of scale factor –0.25, centre (1, 1)
    (f) A rotation of 180° about (2, 10)

Module 29 – Circles

1. **(a)** $y = 110°$, $x = 140°$ **(b)** $q = 32°$, $p = 32°$
    (c) $a = 130°$, $b = 115°$ **(d)** $k = 50°$, $l = 25°$

Module 30 – Properties of 3D Shapes

1. **(a)** 2 faces, 1 edge, 0 vertices
    (b) 7 faces, 15 edges, 10 vertices
    (c) 6 faces, 10 edges, 6 vertices
2. Any suitable answer, e.g.

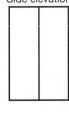

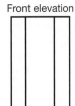

3. A sphere

Module 31 – Perimeter, Area and Volume
1. **(a)** 34m, 28m² **(b)** 42cm, 114cm²
   (c) 4.8π or 15.1m (3 s.f.), $\frac{144}{25}$ π or 18.1m² (3 s.f.)
2. **(a)** 600cm³ **(b)** 1500π or 4710cm³ (3 s.f.)
   (c) 504π or 1583cm³ (4 s.f.)
3. $\frac{40}{3}$ π + 15 or 56.9cm (3 s.f.); 50π or 157cm² (3 s.f.)

Module 32 – Pythagoras' Theorem and Trigonometry
1. Yes. 4.2² + 5.6² = 7² Pythagoras' theorem holds therefore right-angled.
2. **(a)** 40.6° (3 s.f.) **(b)** 58.4° (3 s.f.) **(c)** 58.2° (3 s.f.)
3. **(a)** 6.11 (3 s.f.) **(b)** 3.66 (3 s.f.) **(c)** 9.61 (3 s.f.)

Module 33 – Vectors
1. **(a)** $\begin{pmatrix} 7 \\ -2 \end{pmatrix}$ **(b)** $\begin{pmatrix} 9 \\ 2 \end{pmatrix}$ **(c)** $\begin{pmatrix} 25 \\ 1 \end{pmatrix}$ **(d)** $\begin{pmatrix} 7 \\ 1 \end{pmatrix}$
2. $\overrightarrow{OP} = \frac{1}{2}$ **a**, $\overrightarrow{AB} = -$**a** + **b**, $\overrightarrow{CR} = \frac{1}{2}$ (**b** – **c**)

Module 34 – Experimental and Theoretical Probability
1. **(a)** $\frac{3}{6}$ or $\frac{1}{2}$
   (b) $\frac{3}{6}$ or $\frac{1}{2}$
   (c) $\frac{4}{6}$ or $\frac{2}{3}$
2. **(a)** 0.54 **(b)** 230
3. **(a)** Max, because he has done more trials.
   (b) $\frac{47}{70}$

Module 35 – Representing Probability
1. **(a)** $\frac{1}{36}$ **(b)** $\frac{30}{36}$ or $\frac{5}{6}$ **(c)** 7
2. $\frac{18}{42}$ or $\frac{3}{7}$
3. $\frac{14}{20}$ or $\frac{7}{10}$

Module 36 – Data and Averages
1. Because it is assumed each value is at the midpoint for each group when calculating the mean.
2. No. Needs to use a bigger sample and ask a variety of boys and girls of different ages.
3. 10.53 train, 3h 25mins

Module 37 – Statistical Diagrams
1. **(a)** Any suitable answer, e.g. Slowdown in people's spending, recession, increases in prices/tax.
   (b) On an upward trend, so anything in the range 2.6–3 million.

Module 38 – Comparing Distributions
1. **(a)** 0.2, 0.8, 3.3, 5.6, 4.8, 3.5, 1.9
   (b) Find 50 minutes on the x-axis and read approx. 120 on the y-axis. So approx. 80 people took longer.
2. **(a)** Box plot drawn with minimum = £26, LQ = £55, median = £66, UQ = £77, maximum = £90
   (b) Compare the median and the IQR. Median spending for second survey was lower so they spent less but the variation was greater so there is a much bigger spread of the amount people spent.

Exam Practice Questions
Number
1. **(a)** 4 × 58 = 232 **[1]** 23.2 **[1]**
   (b) 23 × 427 = 9821 **[2]** (use long multiplication) 98.21 **[1]**
   (c) 2400 ÷ 8 **[1]** (multiply both figures by 100) = 300 **[1]**
   (d) 468 ÷ 36 **[1]** (multiply both figures by 10, then long division) = 13 **[1]**
   (e) –7 **[1]**
   (f) –3 **[1]**
   (g) –5 **[1]**
   (h) –140 **[1]**
   (i) 4 + 63 = 67 **[1]**
   (j) –31 – 9 **[1]** = –40 **[1]**
2. **(a)** 36 = 2 × 2 × 3 × 3 **[1]** (use factor trees) = $2^2 × 3^2$ **[1]**
   (b) 48 = 2 × 2 × 2 × 2 × 3 **[1]** (use factor trees) = $2^4 × 3$ **[1]**
   (c) HCF = 2 × 2 × 3 **[1]** (the common prime factors) = 12 **[1]**
   (d) LCM = 12 × 2 × 2 × 3 **[1]** (HCF multiplied by the remaining prime factors) = 144 **[1]**
3. **(a)** $\frac{4 \times 5}{3 \times 6}$ **[1]** $= \frac{20}{18} = \frac{10}{9} = 1\frac{1}{9}$ **[1]**
   (b) $7\frac{11}{12} - (2\frac{3}{4} + 3\frac{5}{6})$ **[1]** $= 7\frac{11}{12} - (6\frac{7}{12})$ **[1]** $= 1\frac{1}{3}$ **[1]**
   (c) $\frac{16}{5} \times \frac{4}{9}$ **[1]** $= \frac{64}{45}$ **[1]** (convert to improper fractions, flip second fraction and multiply) $= 1\frac{19}{45}$ **[1]**
4. **(a)** $5^{3--5} = 5^8$ **[1]**
   (b) $\sqrt{49} \times \sqrt{2} = 7\sqrt{2}$ **[1]**
   (c) $4 \times \sqrt{7} - 3 \times (\sqrt{7})^2 = 4 \times \sqrt{7} - 3 \times 7$ **[1]** $= 4\sqrt{7} - 21$ **[1]**
5. **(a)** 1 **[1]**
   (b) $\frac{1}{9^{\frac{1}{2}}} = \frac{1}{\sqrt{9}}$ **[1]** $= \frac{1}{3}$ **[1]**
   (c) $(\sqrt[3]{27})^4 = 3^4$ **[1]** = 81 **[1]**
6. **(a)** **(i)** 4.36×10^4 **[1]** **(ii)** 8.03×10^{-3} **[1]**
   (b) $(1.2 \div 3) \times (10^{8-4})$ **[1]** $= 0.4 \times 10^4 = 4 \times 10^3$ **[1]**
7. **(a)** $x = 0.4\dot{8}$ 100x = 48.4$\dot{8}$ 99x = 48 **[1]** $x = \frac{48}{99}$ **[1]** $= \frac{16}{33}$
   (b) $x = 0.1\dot{2}\dot{3}$ 10x = 1.2$\dot{3}$ 1000x = 123.2$\dot{3}$ 990x = 122 **[1]**
   $x = \frac{122}{990}$ **[1]** $= \frac{61}{495}$
8. **(a)** $\frac{7}{\sqrt{5}} \times \frac{\sqrt{5}}{\sqrt{5}}$ **[1]** $= \frac{7\sqrt{5}}{5}$ **[1]**
   (b) $\frac{3}{1+\sqrt{2}} \times \frac{1-\sqrt{2}}{1-\sqrt{2}} = \frac{3-3\sqrt{2}}{1^2-(\sqrt{2})^2}$ **[1]** $= \frac{3-3\sqrt{2}}{1-2}$
   $= \frac{3-3\sqrt{2}}{-1} = 3\sqrt{2} - 3$ **[1]**
9. Garage lower bound = 5.5m **[1]** is smaller than car upper bound = 5.55m **[1]**. No **[1]** the car will not definitely fit.
10. $\frac{\text{Lower}^a}{\text{Upper}^b} = \frac{4.25}{2.65}$ **[1 for each correct bound]** = 1.604 **[1]**

Algebra
1. $\frac{5}{x-2} - \frac{1}{x-5} = \frac{5(x-5)-(x-2)}{(x-2)(x-5)}$ **[1]** $= \frac{4x-23}{(x-2)(x-5)}$ **[1]**
2. A and C are parallel. **[1]** B and D are perpendicular. **[1]**
3. A translation **[1]**, of 90° to the left **[1]**, and a translation **[1]** 2 units in the direction of the y-axis **[1]**.
4. **(a)** (x + 9)(x + 7) = 0 **[1]** x = –9 and x = –7 **[2]**
   (b) $(x-4)^2 - 6 = 0$ **[2]** $x = 4 \pm \sqrt{6}$ **[1]**
   (c) $x = \frac{5 \pm \sqrt{45}}{2}$ **[1]** x = 5.85 **[1]** and x = –0.85 **[1]**
5. $y(x - 2) = 3x + 1$
   $xy - 2y = 3x + 1$ **[1]**
   $xy - 3x = 2y + 1$
   $x(y - 3) = 2y + 1$ **[1]**
   $x = \frac{2y+1}{y-3}$ **[1]**

6. $a + b + c = 5$
$4a + 2b + c = 21$ **[1]**
$9a + 3b + c = 47$

$5a + b = 26$
$3a + b = 16$ **[1]**

$a = 5, b = 1, c = -1$
$U_n = 5n^2 + n - 1$ **[3]**

7. **(a)** $x^2 + y^2 = 100$ **[2]**

(b) $x^2 + \dfrac{16x^2}{9} = 100$ **[1]**

$25x^2 = 900$ **[1]**
$x = 6$ **[1]**
$y = 8$ **[1]**

(c) $6x + 8y = 100$ **[2]**
(or $3x + 4y = 50$)

Ratio, Proportion and Rates of Change

1. $2.5 \times 100 \times 100$ **[1]** $= 25\,000\,\text{cm}^2$ **[1]**

2. $2400 \div 4$ **[1]** $= 600$
$600 \times (3 + 4 + 5)$ **[1]** $= £7200$ **[1]**
$\dfrac{5}{12}$ **[1]**

3. $77 \div 1.22$ or $£64 \times 1.22$ **[1]**
Game for 77 euros is cheaper **[1]** by 89p or €1.08 **[1]**

4. $\dfrac{824\,000 - 760\,000}{824\,000}$ or $\dfrac{64\,000}{824\,000}$ **[1]** $= 0.078$

7.8% **[1]**

5. **(a)** $45\,000 \times 1.04 = 46\,800$ **[1]** $\rightarrow 48\,672 \rightarrow 50\,618.88$ **[1]**
$\rightarrow 47\,075.56 \rightarrow$ No, now worth £44 722 **[1]**

(b) $1.04^3 \times 0.93 \times 0.95$ **[1]** $= 0.9938$ **[1]**

6. $22.5 \times \sqrt{4} = 45$ **[1]**

$15 = \dfrac{45}{\sqrt{M}}$ **[1]** $\sqrt{M} = 3, M = 9$ **[1]**

7. Scale factor length = 2, so scale factor volume = 2^3 **[1]** $= 8$
$\dfrac{1600}{8}$ **[1]** $= 200, 1600 - 200 = 1400\,\text{cm}^3$ **[1]**

Geometry and Measures

1. Exterior angle $180 - 156$ **[1]** $= 24°$ $\dfrac{360}{24} = 15$ **[1]**

2. Example answer: Angle $BHG = 180 - (46 + 92)$ **[1]**
$= 42°$ (angles on straight line = 180°)
Angle $HGC = 42°$ (alternate angles are equal)
Angle $HGD = 180 - 42 = 138°$ (angles on straight
line = 180°) **[1]**
All reasons correct **[1]**

3. $QS = RS$ and angle Q = angle R = 90° **[1]**
Angle QOS = angle ROS = $(180 - (90 + 42))$ **[1]** $= 48°$
Angle $POQ = 180 - 2 \times 48 = 84°$ **[1]**

4. Angle $SRT = 180 - (20 + 68) = 92°$ **[1]**
Diagram **[1]** ASA so congruent **[1]**

5. Any rotation **[1]**, $(1, -2), (1, -4), (4, -4)$ **[1]**
Correct reflection of the rotated image **[1]** to $(-1, -2)$,
$(-1, -4), (-4, -4)$
Reflection in $y = -x$ **[1]**

6. $\dfrac{\sin 64}{7} = \dfrac{\sin A}{6}$ **[1]**
$\angle A = 50.4°$ **[1]**
$\dfrac{1}{2} \times 7 \times 6 \sin 65.6$ **[1]** $= 19.1\,\text{cm}^2$ **[1]**

7. $NQ = \sqrt{NP^2 - QP^2} = \sqrt{6^2 - 4^2}$ **[1]**

$NQ = \sqrt{20}$ $\sin M = \dfrac{\sqrt{20}}{7}$ **[1]** $= 0.6389$

$M = 39.7°$ **[1]**

Probability

1. **(a)** $\dfrac{2}{11}$ **[1]**

(b) $\dfrac{4}{11}$ **[1]**

(c) $\dfrac{9}{11}$ **[1]** **(d)** 0 **[1]**

2. **(a)** $1 - (0.1 + 0.2 + 0.25 + 0.1 + 0.15)$ **[1]** $= 0.2$ **[1]**
(b) 0.15×60 **[1]** $= 9$ **[1]**

3. **(a)** HH, HT, TH, TT **[1]** $\dfrac{1}{4}$ **[1]**

(b) HH, HT, TH, TT **[1]** $\dfrac{3}{4}$ **[1]**

4. **(a)**
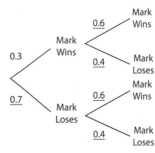

0.7 **[1]** 0.6, 0.4, 0.6, 0.4 **[1]**

(b) 0.3×0.6 **[1]** $= 0.18$ **[1]**
(c) $1 - (0.7 \times 0.4) = 1 - 0.28$ **[1]** $= 0.72$ **[1]**

5.
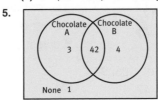

(a) Fully correct with labels **[1]** Two numbers in the correct place **[1]**
(b) 1 identified **[1]** $\dfrac{1}{50}$ **[1]**

(c) $\dfrac{3}{?}$ or $\dfrac{?}{45}$ **[1]** $\dfrac{3}{45}$ or $\dfrac{1}{15}$ **[1]**

6.
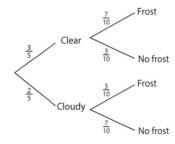

$P(\text{Frost}) = P(\text{Clear/Frost}) + P(\text{Cloudy/Frost})$

$\dfrac{3}{5} \times \dfrac{7}{10} = \dfrac{21}{50}$ **[1]** $\dfrac{2}{5} \times \dfrac{3}{10} = \dfrac{6}{50}$ **[1]**

$\dfrac{21}{50} + \dfrac{6}{50}$ **[1]** $= \dfrac{27}{50}$ **[1]**

Statistics

1. **(a)** Correct midpoints **[1]** Total of 8150 **[1]**
$8150 \div 200$ **[1]** $= 40.8\,\text{kg}$ (3 s.f.) **[1]**
(b) $35 \leqslant w < 45$ **[1]**
(c) $45 \leqslant w < 55$ **[1]**

2. Frequency density on y-axis and weight on x-axis with no gaps and
continuous scale. **[1]**
Correct frequency densities of 7.2, 2.4, 1.3, 3.1, 4, 3.3, 2.3 **[1]**
Bars drawn correctly **[1]**

3. **(a)** Cumulative frequency calculated correctly (36, 60, 73, 104,
144, 177, 200) **[1]**
Points correctly plotted on the upper values for each group at
(15, 36), (25, 60), (35, 73), (45, 104), (55, 144), (65, 177),
(75, 200). **[1]** Points connected together. **[1]**
(b) Line drawn at 60kg **[1]** Answer 37–40 **[1]**

4. 3, 3, 5, 6, 13. So greatest range = 10 **[3]**

GCSE
Success

Maths

Higher Tier

Exam Practice Workbook

Deborah Dobson,
Phil Duxbury,
Mike Fawcett and
Aftab Ilahi

Number

Algebra

Ratio, Proportion and Rates of Change

Contents

Geometry and Measures

Probability

Statistics

Practice Exam

1 **(a)** Work out −4 + −3 🖩 [1 mark]

....

(b) Work out (−3 × −2) + (−5 × −3) 🖩 [1 mark]

....

(c) Work out (−8 − −5) × (7 − 12) ÷ (−2 − 3) 🖩 [1 mark]

....

2 Are the following **true** or **false**?

(a) −3 + −2 < −2 − +3 .... [1 mark]

(b) −5 − −7 = 11 + −9 .... [1 mark]

(c) −4 × −7 × −10 = 10 × −4 × 7 .... [1 mark]

(d) 26 ÷ 15 > −28 ÷ −15 .... [1 mark]

3 The temperature in New York is 8°C. It is 11°C colder in London.

(a) What is the temperature in London? [1 mark]

....

Moscow is three times colder than London.

(b) Find the temperature in Moscow. [1 mark]

....

4 Given that 129 × 42 = 5418, find the following. 🖩

(a) 1.29 × 4.2 .... [1 mark]

(b) 12.9 × 420 .... [1 mark]

(c) 5418 ÷ 0.42 .... [1 mark]

(d) 54.18 ÷ 0.129 .... [1 mark]

5 Bethany has £40. Tomorrow she will be paid £150 but will pay £225 for a holiday.

(a) How much debt will Bethany be in? [1 mark]

....

Bethany's father gives her $\frac{2}{5}$ of the amount she is in debt.

(b) How much debt is Bethany in now? [2 marks]

....

Score /16

For more help on this topic, see Letts GCSE Maths Higher Revision Guide pages 4–5.

1 Write down all the factors of 48 which are

    (a) odd numbers   [1 mark]

    ....

    (b) square numbers   [1 mark]

    ....

    (c) multiples of 6   [1 mark]

    ....

2 James is thinking of a number between 1 and 50. It is 4 more than a prime number and 4 less than a square number. It is also a multiple of 3.

    What could James's number be?   [2 marks]

                                   .... or ....

3 **(a)** Find the lowest common multiple (LCM) of 5 and 11.   [1 mark]

                                   ....

    (b) Find the highest common factor (HCF) of 14 and 35.   [1 mark]

                                     ....

4 **(a)** Write 72 as a product of its prime factors. Give your answer in index form.   [3 marks]

                                     ....

    (b) Write 90 as a product of its prime factors. Give your answer in index form.   [3 marks]

                                     ....

    (c) Find the highest common factor (HCF) of 72 and 90.   [2 marks]

                                     ....

    (d) Find the lowest common multiple (LCM) of 72 and 90.   [2 marks]

                                     ....

5 Bus A leaves the bus station every 12 minutes.
Bus B leaves the bus station every 28 minutes.
Bus A and Bus B both leave the bus station at 8.00am.

    When will they next both leave the bus station at the same time?   [3 marks]

                                     ....

Score   **/20**

Factors, Multiples and Primes

Module 2

For more help on this topic, see Letts GCSE Maths Higher Revision Guide pages 6–7.

1 (a) Work out $5 + 4^2 \times 3$ 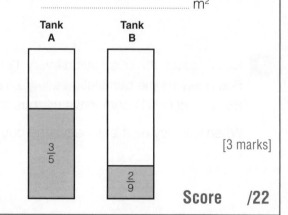 [1 mark]

....

(b) Work out $\frac{1}{3} + \frac{2}{5}$ [1 mark]

....

(c) Write down the reciprocal of 0.25 [1 mark]

....

2 Work out $\dfrac{(8 - 5 \times 2)^2}{5 - 3}$ .... [2 marks]

3 Calculate the following. Give your answers in their simplest form.

(a) $\frac{4}{5} \times \frac{3}{8}$ .... [2 marks]

(b) $\frac{5}{6} \div \frac{3}{10}$ .... [2 marks]

(c) $3\frac{3}{4} - 2\frac{3}{5}$ .... [2 marks]

(d) $1\frac{1}{2} \div 3\frac{4}{5}$ .... [2 marks]

4 Wes is drawing storyboards for a new film. One day he completes $\frac{2}{5}$ of the drawings.

On the next day he completes $\frac{3}{7}$ of the drawings.

What fraction of the drawings has he got left to complete? [3 marks]

....

5 A wall measures $5\frac{1}{4}$m by $3\frac{5}{6}$m.

Find the area of the wall, giving your answer as a mixed number. [3 marks]

.... m²

6 Tank A and Tank B are identical water tanks.

Tank A is $\frac{3}{5}$ full. Tank B is $\frac{2}{9}$ full.

Tank B is emptied into Tank A.

What fraction of Tank A is now empty? [3 marks]

....

Score /22

For more help on this topic, see Letts GCSE Maths Higher Revision Guide pages 8–9.

1 Write the following in order of value, smallest first.

2^5 $\qquad$ $\sqrt[3]{64}$ $\qquad$ 3^3 $\qquad$ 5^2 $\qquad$ $\sqrt[4]{625}$ [1 mark]

....

2 Find the value of x in each of the following.

(a) $2^x = 128$ .... [1 mark]

(b) $\sqrt[3]{x} = 5$ .... [1 mark]

(c) $x^4 = 81$ .... [1 mark]

3 **(a)** Write $4^6 \div 4^{-4}$ as a power of 4 in its simplest form. 🖩 [1 mark]

....

(b) Write $\sqrt{32}$ as a surd in its simplest form. 🖩 [1 mark]

....

(c) Write 3^{-2} as a fraction in its simplest form. 🖩 [1 mark]

....

4 **(a)** Write the following numbers in standard form. 🖩

(i) 321 000 .... [1 mark]

(ii) 0.000 605 .... [1 mark]

(b) Calculate the following. Leave your answer in standard form. 🖩

(i) $(5 \times 10^3) \times (6 \times 10^{-8})$ .... [2 marks]

(ii) $(4.1 \times 10^4) + (3.4 \times 10^3)$ .... [2 marks]

5 Simplify the following. Leave your answer as a surd in its simplest form. 🖩

(a) $\sqrt{3}(4 - \sqrt{3})$ .... [1 mark]

(b) $(2 + 3\sqrt{5})(4 - 2\sqrt{5})$ .... [2 marks]

6 Rationalise the denominator of the following surds. 🖩

(a) $\dfrac{4}{\sqrt{7}}$ .... [1 mark]

....

(b) $\dfrac{5}{2 - \sqrt{3}}$ [3 marks]

....

Score /20

Powers and Roots

Module 4

For more help on this topic, see Letts GCSE Maths Higher Revision Guide pages 10–11.

1 Write the following in order of value, smallest first. 🔲

$\frac{12}{25}$ $\frac{3}{8}$ 0.475 $\frac{17}{51}$

[1 mark]

2 Write 47.5% as a fraction in its simplest form.

[1 mark]

3 Which of the fractions $\frac{5}{8}$ and $\frac{7}{11}$ is closest to $\frac{2}{3}$? Show your working. 🔲

[3 marks]

4 Calculate the following. 🔲

(a) 3.7 × 4.9

[2 marks]

(b) 369.6 ÷ 1.4

[2 marks]

5 Carl orders 34 Creamy Crisp Doughnuts for a church event.

Each doughnut costs £1.29

Find the total cost of the doughnuts. Show your working. 🔲

[3 marks]

6 Write $0.\dot{4}$ as a fraction in its simplest form.

[1 mark]

7 Prove that $0.\dot{4}\dot{5}$ is equivalent to $\frac{5}{11}$

[2 marks]

8 Write $0.5\dot{3}\dot{6}$ as a fraction in its simplest form.

[2 marks]

Score   /17

For more help on this topic, see Letts GCSE Maths Higher Revision Guide pages 12–13.

1 Evaluate the following numbers to the degree of accuracy shown.

(a) $\sqrt{90}$ (2 decimal places)   ....   [1 mark]

(b) 2.56^2 (3 decimal places)   ....   [1 mark]

(c) $\sqrt[3]{175.6}$ (2 decimal places)   ....   [1 mark]

2 (a) Round 404 928 to 2 significant figures.   [1 mark]

....

(b) Round 0.04965 to 3 significant figures.   [1 mark]

....

(c) Round π to 4 significant figures.   [1 mark]

....

3 Violet runs a coffee lounge. She wants to make 35 cups of coffee from a 500g bag of coffee.

How much coffee should she use for each cup? Give your answer to 3 significant figures.

.... g   [2 marks]

4 Rachel works in a school office. She sends, on average, 72 letters home each week. Each letter costs 59p to send. There are 39 weeks in the school year.

Estimate the cost of sending the letters home for the whole year.   🔄   [3 marks]

£ ....

5 Ethan is 1.73m tall to the nearest centimetre.

Write down the upper and lower bounds for his height.   [2 marks]

Upper bound: ....m   Lower bound: ....m

6 A water bottle holds 500ml of water to the nearest millilitre.

What is the maximum amount of water in a pack of 8 bottles?   [2 marks]

.... ml

7 The volume of a shape can be found using the formula $v = \frac{m}{d}$, where m is mass and d is density.

If $m = 6.8$ to 2 significant figures and $d = 0.34$ to 2 significant figures, find the minimum and maximum value for v to 2 decimal places.   [4 marks]

Minimum value: ....   Maximum value: ....

Score   /19

Approximations

Module 6

1 Calculate the following. Write down all the figures on your calculator display.

(a) 4.5^4 .... [1 mark]

(b) $\sqrt[4]{2401}$ .... [1 mark]

(c) $\sqrt[7]{78125}$ .... [1 mark]

2 Calculate $\dfrac{4 + 3.9^2}{5.2 \times 6.3}$

Write down all the figures on your calculator display. [2 marks]

....

3 Work out $(4.56 \times 10^{-2}) \div (3.8 \times 10^7)$, giving your answer in standard form. [1 mark]

....

4 Calculate the following using the fraction button on your calculator. Give your answer as a mixed number.

(a) $14\dfrac{2}{3} - 11\dfrac{5}{7}$ .... [1 mark]

(b) $4\dfrac{3}{10} \times 7\dfrac{2}{9}$ .... [1 mark]

5 The Sun is 400 times further away from the Earth than the Moon.

The Sun is 1.496×10^8 km away from the Earth.

(a) How far away is the Moon from the Earth? Write your answer in standard form. [3 marks]

.... km

The Moon has a diameter of 3.48×10^3 km. The Sun has a diameter of 1.392×10^6 km.

(b) How many times bigger is the Sun's diameter compared with the Moon's diameter? Write your answer in standard form. [3 marks]

....

(c) Explain why the Sun and the Moon appear the same size in the sky. [1 mark]

....

....

(d) Find the distance between the Sun and the Moon when the Earth is directly between them. Write your answer in standard form to 2 significant figures. [3 marks]

.... km

Score /18

For more help on this topic, see Letts GCSE Maths Higher Revision Guide pages 16–17.

1 Simplify the following expressions. 🖩

    (a) $8a + 2 - 3a$ .... [1 mark]

    (b) $6h + 3k + 9h - k$ .... [1 mark]

    (c) $5a + 2b + 3c - 4a - 3a - 3c$ .... [1 mark]

    (d) $3x^2 - 2x + 5x^2 + x + 1$ .... [1 mark]

2 Simplify the following expressions. 🖩

    (a) $q^2 p \div pq$ .... [1 mark]

    (b) $\left(3p^4\right)^3$ .... [1 mark]

    (c) $18f^7 g^2 \div 3f^2 g$ .... [1 mark]

    (d) $\left(2p^2\right)^{-3}$ .... [1 mark]

3 Simplify the following expressions where possible. 🖩

    (a) $60k^2 \div 5$ .... [1 mark]

    (b) $\sqrt[4]{a^{12}}$ .... [1 mark]

    (c) $\sqrt[3]{p^6} \times \sqrt[6]{p^3}$ .... [1 mark]

    (d) $p^2 \times p^3 \times \sqrt{p^3}$ .... [1 mark]

4 Evaluate 🖩

    (a) 4^{-3} .... [1 mark]

    (b) $\left(\dfrac{2}{5}\right)^{-2}$ .... [1 mark]

    (c) $32^{\frac{2}{5}}$ .... [1 mark]

    (d) 16^0 .... [1 mark]

5 Which of the following are **equations** and which are **identities**? 🖩

    (a) $3x - 4 = (3x + 1) - (x - 5)$ .... [1 mark]

    (b) $\dfrac{p+q}{pq} = \dfrac{1}{p} + \dfrac{1}{q}$ .... [1 mark]

    (c) $x^2 + x^2 = 2x^2$ .... [1 mark]

    (d) $x^2 + x^2 = x^4$ .... [1 mark]

    (e) $\sqrt{a + b} = \sqrt{a} + \sqrt{b}$ .... [1 mark]

Score   /21

For more help on this topic, see Letts GCSE Maths Higher Revision Guide pages 20–21.

1 Expand and simplify $(2x-5y)(4x+3y)$ 📝 [2 marks]

2 Expand and simplify $p(2p+1)(2p-1)$ 📝 [2 marks]

3 Write $3(2x-1)+4(x+8)+5$ in the form $a(bx+c)$ where a, b and c are integers. 📝 [3 marks]

4 Factorise $a^2-19a+48$ 📝 [2 marks]

5 Simplify the expression $\dfrac{x^2-7x-18}{x^2-81}$ 📝 [3 marks]

6 Write $\dfrac{1}{a+2}-\dfrac{1}{a+3}$ as a single fraction. 📝 [2 marks]

7 Simplify $\dfrac{x+4}{x-2}\times\dfrac{2-x}{x+1}$ 📝 [2 marks]

8 Expand and simplify $(x+2)(x-3)(x+4)$ 📝 [4 marks]

9 Simplify the expression $\dfrac{x-9}{6}\div\dfrac{x^2-9x}{3}$ 📝 [4 marks]

10 Factorise $3x^2+19x-40$ 📝 [2 marks]

11 Expand and simplify $(2x+1)^3$ 📝 [3 marks]

12 Simplify the expression $\dfrac{x^2-5x-84}{x^2+5x-14}$ 📝 [3 marks]

Score   **/32**

For more help on this topic, see Letts GCSE Maths Higher Revision Guide pages 22–23.

1 If $x = 4$, $y = -2$ and $z = \dfrac{1}{3}$, find the values of 📝

(a) $(x - y)^2$ [1 mark]

....

(b) $\dfrac{x^2}{z}$ [1 mark]

....

(c) $2z - y$ [1 mark]

....

(d) $xy + z$ [1 mark]

....

(e) $\sqrt{10x + 2y}$ [1 mark]

....

2 Rearrange $y = \dfrac{2 + 3x}{x - 8}$ to make x the subject. 📝 [3 marks]

....

3 Rearrange $s = ut + \dfrac{1}{2}at^2$ to make a the subject. 📝 [2 marks]

....

4 Rearrange $y = \dfrac{2}{x} + 5$ to make x the subject. 📝 [2 marks]

....

5 A suitcase weighs p grams when empty. When it is full of q grams of paper, it weighs r grams. Write down an equation for q in terms of p and r. 📝 [2 marks]

....

6 The formula to convert degrees Fahrenheit to degrees Celsius is $T_C = \dfrac{5(T_F - 32)}{9}$

(a) Use this formula to convert 88 degrees Fahrenheit (T_F) to degrees Celsius. [1 mark]

....

(b) Rearrange this formula to make T_F the subject. [2 marks]

....

(c) Hence convert 60 degrees Celsius to degrees Fahrenheit. [1 mark]

....

Score /18

For more help on this topic, see Letts GCSE Maths Higher Revision Guide pages 24–25.

Algebraic Formulae

Module 10

1 Solve the equation $5(x-2)-3(x-4)=4$ 📱 [2 marks]

....

2 Solve, by factorisation, the equation $x^2-3x=28$ 📱 [3 marks]

....

3 Solve, by factorisation, the equation $3x^2+x-10=0$ 📱 [3 marks]

....

4 $4x-3y=14$ and $x+y=7$, find the values of x and y. 📱 [2 marks]

$x =$ .... $y =$ ....

5 Solve the equation $x^2-8x-4=0$ by completing the square.
Give your answers in surd form. 📱 [3 marks]

....

6 Write $3x^2+24x+40$ in the form $a(x+b)^2+c$ 📱 [5 marks]

....

7 Solve the equation $3x^2-8x-13=0$ by using the quadratic formula.
Give your answers to 3 significant figures. [3 marks]

....

8 C is the curve with equation $y=x^2-x-2$ and L is the line with equation $y=2x+2$

If L and C intersect at two points A and B, find the exact length of AB. 📱 [5 marks]

....

9 Solve the simultaneous equations $y=8-x^2$ and $y=x-4$ 📱 [6 marks]

....

10 If you are given a graph of $y=2x^2-10x+7$, what straight line should you plot
in order to solve (graphically) the equation $4x^2-22x+9=0$? 📱 [4 marks]

....

11 $x^3-7x+2=0$ may be rearranged to give $x=\sqrt[3]{7x-2}$

By taking a starting value of $x_0=1$, use the iteration $x_{n+1}=\sqrt[3]{7x_n-2}$ to find a
solution to $x^3-7x+2=0$ to 2 decimal places. [3 marks]

....

Score /39

For more help on this topic, see Letts GCSE Maths Higher Revision Guide pages 26–27.

1 Solve the inequality $7x - 2 > 33$ 📠 [2 marks]

....

2 State the inequalities as shown on the number lines below. 📝

(a) .... [1 mark]

-7 -6 -5 -4 -3 -2 -1 0 1 2 3 4

(b) .... [1 mark]

-4 -3 -2 -1 0 1 2 3 4 5 6 7

(c) .... [1 mark]

-6 -5 -4 -3 -2 -1 0 1 2 3 4 5 6

3 List all integers x for which $x^2 < 10$ 📠 [2 marks]

....

4 Solve the inequality $x^2 - 9x + 8 \leq 0$, writing your answer using set notation. 📝 [3 marks]

....

5 (a) On the grid below, draw sketches of $y = x^2$ and $y = x$ 📝 [2 marks]

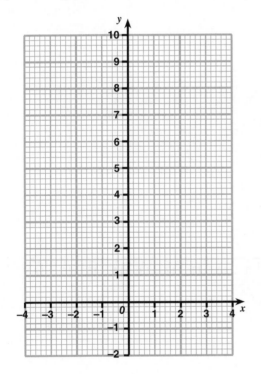

(b) For what values of x is $x^2 < x$? 📝 [1 mark]

....

(c) What are the solutions to $x^2 > x$ in set notation? 📝 [2 marks]

....

6 Draw a rectangle with coordinates (2, 1), (5, 1), (2, 6) and (5, 6), joining the points with straight lines. Shade the rectangle.

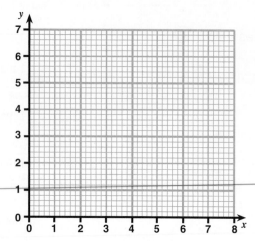

Describe the shaded area using inequalities. [4 marks]

7 Shade the region corresponding to the following inequalities. [4 marks]

$$y < x \qquad x + 3y < 15 \qquad y > 1$$

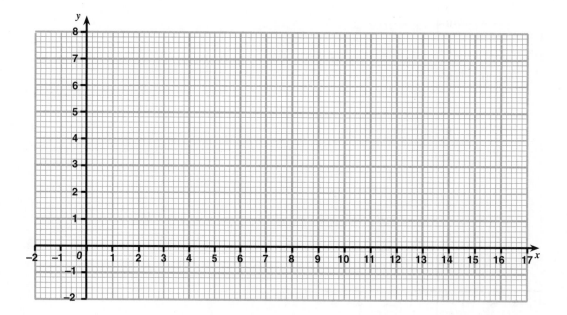

For more help on this topic, see Letts GCSE Maths Higher Revision Guide pages 28–29.

1 Here are the first four terms of an arithmetic sequence.

41   32   23   14

Write down the next two terms in the sequence.   [2 marks]

....

2 Here are the first four terms of a quadratic sequence.

12   17   24   33

Write down the next two terms in the sequence.   [2 marks]

....

3 A term-to-term sequence is given by $U_{n+1} = 5U_n - 2$ where $U_1 = 1$

Write down the first five terms in this sequence.   [2 marks]

....

4 11, 15, 19, 23, … are the first four terms in an arithmetic sequence.

(a) Find an expression for U_n, the position-to-term formula.   [2 marks]

....

(b) Is 303 a number in this sequence? Explain your answer.   [2 marks]

....

5 7, 6, 7, 10, 15, … are the first five terms of a quadratic sequence.

(a) If the position-to-term sequence U_n is given by $U_n = an^2 + bn + c$, find the values of a, b and c.   [5 marks]

$a =$ .... $b =$ .... $c =$ ....

(b) Find U_{20}, the 20th term in the sequence.   [1 mark]

....

6 A geometric sequence has first term $a = 2\sqrt{3}$ and common ratio $r = 2\sqrt{3}$

(a) Write down an expression for the position-to-term formula, U_n [2 marks]

....

(b) Find an exact value for U_3, the third number in the sequence.   [1 mark]

....

Score   **/19**

For more help on this topic, see Letts GCSE Maths Higher Revision Guide pages 30–31.

1 Find the gradients of the following lines labelled A, B, C and D. 📝 [4 marks]

A: $y = 6x - 7$

B: $y = 8 - 3x$

C: $2x + 3y + 4 = 0$

D: $y - x = 0$

Line A: Gradient is ....

Line B: Gradient is ....

Line C: Gradient is ....

Line D: Gradient is ....

2 From the following lines, write down a pair that is parallel and a pair that is perpendicular. 📝 [2 marks]

A: $3y = x - 6$

B: $4x + 2y = 8$

C: $x = 3y - 12$

D: $2y - x + 2 = 0$

Lines .... and .... are parallel.

Lines .... and .... are perpendicular.

3 Write down the equation of the y-axis. [1 mark]

....

4 Write down the equation of the line parallel to $y = 5 - 6x$ that passes through the point (0, 9). [2 marks]

....

5 Work out the equation of the line parallel to $y = 5x - 2$ that passes through the point (−3, 1). [3 marks]

....

6 Find the equation of the line joining points (−4, −10) and (6, −5). [4 marks]

....

7 Find the equation of the line perpendicular to $3y = x + 15$ that passes through the point (5, −4). [5 marks]

....

Score /21

For more help on this topic, see Letts GCSE Maths Higher Revision Guide pages 32–33.

1 State the equation of the line of symmetry of the curve $y = 5x^2 - 30x + 12$ 🖲 [2 marks]

....

2 Find (using algebra) the roots of the equation $x^2 - 7x - 44 = 0$ 🖲 [3 marks]

....

3 Find (using algebra) the roots of the equation $3x^2 + 10x - 8 = 0$ 🖲 [3 marks]

....

4 Consider the curve $y = x^2 + 6x + 4$ 🖲

(a) By completing the square, find the coordinates of the minimum point on the curve. [3 marks]

....

(b) State the coordinates where the curve crosses the y-axis. [1 mark]

....

5 Consider the curve $y = (x - 2)(x - 4) - 48$ 🖲

(a) Find where the curve crosses the x-axis. [4 marks]

....

(b) State the coordinates where the curve intercepts the y-axis. [1 mark]

....

(c) Find the equation of the line of symmetry. [1 mark]

....

6 **(a)** By completing the square, write $3x^2 - 12x + 22$ in the form $p(x + q)^2 + r$ 🖲 [5 marks]

....

(b) Hence find the coordinates of the minimum point on the curve
$y = 3x^2 - 12x + 22$ 🖲 [1 mark]

....

7 By completing the square, find the maximum point on the curve
$y = 25 + 10x - x^2$ 🖲 [4 marks]

....

Score /28

For more help on this topic, see Letts GCSE Maths Higher Revision Guide pages 34–35.

1 Given $f(x) = 5x + 3$, find expressions for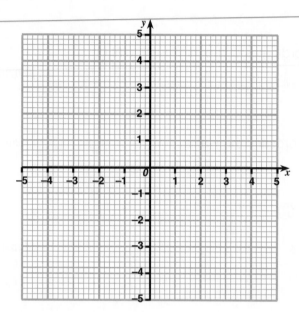

(a) $f^{-1}(x)$ [2 marks]

....

(b) $ff(x)$ [2 marks]

....

2 Sketch the graphs of $y = \dfrac{1}{x}$ and $y = \dfrac{-4}{x}$ on the same axes. [4 marks]

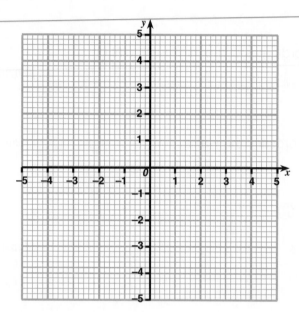

3 Given $f(x) = \dfrac{1}{x}$ and $g(x) = 4x - 6$, find

(a) $g\left(\dfrac{1}{2}\right)$ [1 mark]

....

(b) $f\left(\dfrac{-1}{3}\right)$ [1 mark]

....

(c) $fg(1) - gf(1)$ [2 marks]

....

4 On the grid on page 21, sketch graphs of

(a) $y = 2^x$ [2 marks]

(b) $y = 2^x + 3$ [2 marks]

(c) $y = 2^{x+3}$ [2 marks]

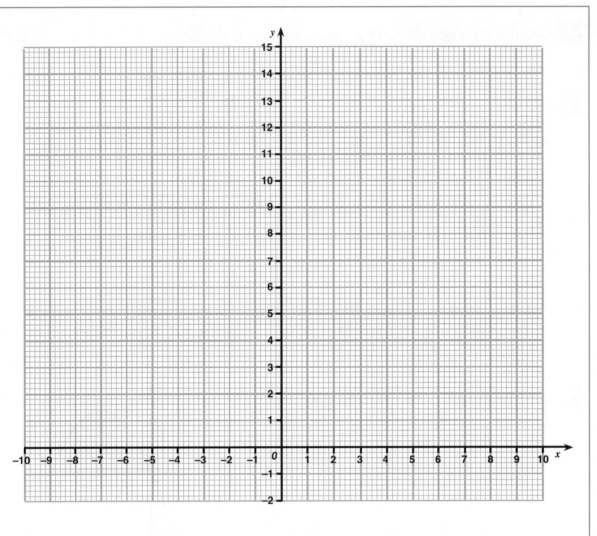

5 The graph of $y = \cos x$ transforms to $y = 2 + \cos(x - 90°)$ through a sequence of two transformations. Describe each transformation. 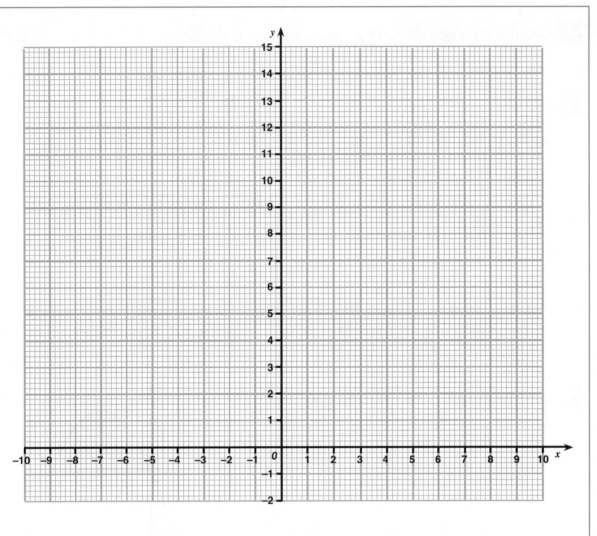 [4 marks]

....

....

....

6 Given $p(x) = \dfrac{3x+1}{1-x}$, find a simplified expression for $p^{-1}(x)$ [5 marks]

....

7 The graph of $y = x^2 - 2x + 5$ is found after applying two transformations to the graph of $y = x^2$

By writing $x^2 - 2x + 5$ in the form $(x - a)^2 + b$, describe the two transformations. [6 marks]

....

....

Score   **/33**

For more help on this topic, see Letts GCSE Maths Higher Revision Guide pages 36–37.

Other Functions

Module 16

1 Describe the shape defined by the equation $x^2 + y^2 = 2$ 🖉 [3 marks]

....

2 Consider the circle having equation $4x^2 + 4y^2 = 9$ 🖉

(a) Sketch the circle. [3 marks]

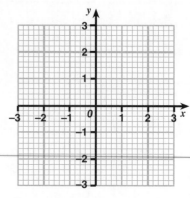

(b) Find an equation for the tangent to the circle at the point $\left(\frac{1}{2}, \sqrt{2}\right)$, giving your answer in the form $ax + by + c = 0$ [4 marks]

....

3 Consider the graph of $y = 12x - 4x^2$ shown.

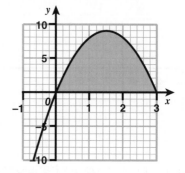

(a) Estimate the area bounded by the curve and the x-axis by splitting the area into six strips. [6 marks]

.... units^2

(b) Find an estimate for the equation of the tangent to the curve at the origin. [5 marks]

....

4 Given a circle $x^2 + y^2 = a^2$, prove that the equation of the tangent at the point (h, k) lying on the circumference is given by $hx + ky = a^2$ 🖉 [6 marks]

....

....

....

....

....

Score /27

For more help on this topic, see Letts GCSE Maths Higher Revision Guide pages 38–39.

1 Calculate the area of a rectangle with width 850mm and length 1.25m. [2 marks]

....

2 Convert 3.4km² to metres². 🖩 [2 marks]

....

3 If 1cm³ of water has a mass of 1g, what is the mass of 1m³ of water in kilograms? 🖩 [2 marks]

....

4 5 miles = 8km

Which is faster, 68mph or 110km/h? [2 marks]

....

5 Convert 24m/s to kilometres per hour. [3 marks]

....

6 Shahida works 30 hours each week and earns £18.40/h.

Bob works 35 hours each week and earns £626.00.

Leila works 32 hours each week and earns £30000 per annum.

(a) Who has the best hourly rate? [3 marks]

....

(b) Who earns the most per year? [2 marks]

....

7 Here are two pieces of gold:

Block A has a volume of 24cm³ and a mass of 463g.

The volume of block B is 30cm³ and it contains exactly the same quality of gold.

A   B

(a) Calculate the density of gold. [2 marks]

....

(b) Calculate the mass of block B. [2 marks]

....

(c) If the price of gold is £23.50/g, calculate the total value of both pieces. [2 marks]

....

Score /22

For more help on this topic, see Letts GCSE Maths Higher Revision Guide pages 42–43.

1 A model plane is made using a scale of 1:20 🔲

    (a) If the wingspan of the model is 3m, what is the wingspan of the actual plane? [1 mark]

....

    (b) If the plane is 57m long, how long is the model? [2 marks]

....

2 Here is a map with a scale of 1:5 000 000 🔲

    (a) What is the distance on the map between Bristol and Leeds? [1 mark]

....

    (b) What is the actual distance? [2 marks]

....

    (c) Measure the bearing of Exeter from Leeds. .... [2 marks]

    (d) What is the bearing of Leeds from Exeter? .... [2 marks]

3 A cruise ship sets off from harbour A for 100km on a bearing of 075° to point B. It then changes to a bearing of 220° and sails at 15km/h for 6 hours to point C. Finally it heads straight back to port.

(a) Draw an accurate diagram to show the journey using a scale of 1cm:20km [2 marks]

(b) Measure the bearing and distance of the last leg of the journey from C to A. [2 marks]

Bearing: .... Distance: .... km

4 A map is drawn to a scale of 1:25000

Villages S and R are 18cm apart on the map.

(a) Work out the actual distance between S and R in kilometres. [2 marks]

....

R is due West of S. T is another village. The bearing of T from R is 040° and the bearing of T from S is 320°.

(b) Draw a diagram to show the positions of S, R and T. [2 marks]

Score /18

For more help on this topic, see Letts GCSE Maths Higher Revision Guide pages 44–45.

1 Circle the numbers which are **not** equivalent to 3.75 🗐 [2 marks]

$$\frac{30}{8} \qquad 37.5\% \qquad 375\% \qquad \frac{14}{5} \qquad 3\frac{5}{8} \qquad \frac{750}{200}$$

2 Write the following numbers in ascending order. 🗐 [2 marks]

$$0.21 \qquad 20\% \qquad \frac{3}{10} \qquad 0.211 \qquad \frac{2}{9}$$

3 Is 2.125 or $2\frac{4}{5}$ closer to $2\frac{1}{2}$? Explain your reasoning. 🗐 [2 marks]

4 On a new estate of 32 houses, $\frac{3}{8}$ have two bedrooms. $\frac{5}{6}$ of the two-bedroom houses have a garage.

What percentage of the whole estate is represented by two-bedroom houses with a garage? [2 marks]

5 Claire makes soft toys to sell at a Christmas market.

(a) Each dog toy costs £3.45 to make and Claire sells them for £5.99.

What is her percentage profit? [2 marks]

(b) A tiger toy costs 15% more to make than a dog toy and she makes 80 tiger toys. Claire sells 55 of them for £6.99 and the rest at the reduced price of £4.

What percentage profit does she make on tiger toys? [4 marks]

6 At Mathstown School 55% of the students are girls. 40% of the girls and 65% of the boys have school lunch.

(a) What percentage of students at the school have school lunch? [3 marks]

(b) What fraction of the boys do not have school lunch? [2 marks]

Score /19

For more help on this topic, see Letts GCSE Maths Higher Revision Guide pages 46–47.

1 Simplify these ratios and circle the odd one out. You must show all your working. 🖉 [2 marks]

£4 : £6   10 : 15   20cm : 3m   750g : 1.125kg   40 seconds : 1 minute

....

....

....

2 Jane is making 'mist blue' paint for her room. She mixes navy blue, grey and white paint in the ratio 1 : 2 : 7

(a) How much of each colour does Jane need to make 2 litres of paint? [3 marks]

Navy blue: .... ml   Grey: .... ml   White: .... ml

(b) Jane finds she has $\frac{3}{4}$ litres of navy blue, 1200ml of grey and 6 litres of white paint.

What is the maximum amount of 'mist blue' she can make? [2 marks]

.... litres

3 The ratio of A : B is 5 : 8

Complete this statement. A is $\dfrac{\square}{\square}$ of B. 🖉 [1 mark]

4 This is a recipe for shortbread:

Makes 15 biscuits			
110g butter	50g sugar	175g flour	50g chocolate chips

(a) Amil has 70g of sugar. How many biscuits can he make? [2 marks]

....

(b) How much flour is needed to make 12 biscuits? [2 marks]

.... g

5 Lucy makes green paint by mixing yellow and blue paint in the ratio 5 : 2
Blue paint costs £30 for 5 litres and yellow paint costs £28 for 7 litres.

Lucy sells her paint for £4.50 per litre. Will she make a profit? Show your working to justify your decision. 🖉 [3 marks]

....

....

Score   /15

For more help on this topic, see Letts GCSE Maths Higher Revision Guide pages 48–49.

Ratio

Module 21

1 If Shabir has 250ml of soup for her lunch, how many kilocalories of energy will she get? .... [2 marks]

2 Leon changes £500 to euros at the rate shown and goes to France on holiday.

£1 = 1.29 euros	*£1 = 187.99 Japanese yen*
£1 = 1.56 US dollars	*£1 = 97.04 Indian rupees*

(a) How many euros does he take on holiday? .... [1 mark]

Leon spends €570.

(b) He changes his remaining euros on the ferry where the exchange rate is £1 : €1.33

How much in pounds sterling does he take home? .... [2 marks]

3 James' dairy herd of 80 cattle produces 1360 litres of milk per day.

(a) If James buys another 25 cattle and is paid 30p/litre, what will his annual milk income be? .... [4 marks]

(b) If 6 tonnes of hay will last 80 cattle for 10 days, how long will the same amount of hay last the increased herd? [2 marks]

....

4 Triangles *PQR* and *STU* are similar.

Find the missing lengths *PR* and *TU*. [4 marks]

PR = ...

TU = ...

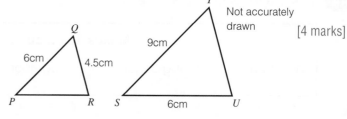

Not accurately drawn

5 Two similar cylinders P and Q have surface areas of 120cm² and 270cm².

If the volume of Q is 2700cm³, what is the volume of P? [3 marks]

Q

Surface area: 270cm²

Volume: 2700cm³

P

Surface area: 120cm²

.... cm³

Score /18

For more help on this topic, see Letts GCSE Maths Higher Revision Guide pages 50–51.

1 (a) Peter invests £10 000 in a savings account which pays 2% compound interest per annum.

How much will his investment be worth after four years? [2 marks]

....

(b) Paul invests £10 000 in company shares.
In the first year the shares increase in value by 15%.
In the second year they increase by 6%.
In the third year they lose 18% of their value.
In the fourth year the shares increase by 1%.

What is his investment worth after four years? .... [3 marks]

2 Lazya invests £6500 at 3% compound interest for three years. She works out the first year's interest to be £195. She tells her family she will earn £585 over three years.

Is she right? Show working to justify your decision. [3 marks]

....

....

3 This graph shows a tank being filled with water.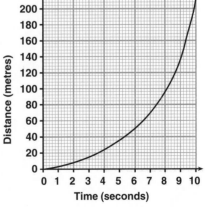

(a) How deep is the water when the tank is full?

.... [1 mark]

(b) Between what times is the tank filling fastest?

.... [1 mark]

(c) Work out the rate of decrease of water level as the tank empties. .... [1 mark]

4 This graph shows the distance travelled by a cyclist for the first 10 seconds of a race.

(a) Work out the cyclist's average speed for the first 10 seconds.

.... [2 marks]

(b) Estimate the actual speed at 5 seconds.

.... [3 marks]

Score /16

For more help on this topic, see Letts GCSE Maths Higher Revision Guide pages 52–53.

1 Using only a ruler and a pair of compasses, construct a triangle with sides 8cm, 7cm and 9cm.

[2 marks]

2 Using only a ruler and a pair of compasses, construct a rectangle with one side 8cm and an area of 28cm².

[2 marks]

3 Using only a ruler and a pair of compasses, draw a line *PQ* such that angle *PQR* is 90°.

[2 marks]

Q ————————————————————————— *R*

4 This is a plan of a field using a scale of 1cm : 10m

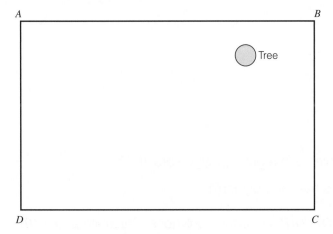

The farmer wants to install a water trough. It must be

- at least 15m from the tree
- more than 20m from the fence *DC*
- nearer to *AB* than to *AD*.

Shade the area where the farmer can put the trough.   [3 marks]

5 Using only a ruler and a pair of compasses, construct an angle of 30°.   [3 marks]

Score   /12

For more help on this topic, see Letts GCSE Maths Higher Revision Guide pages 56–57.

1 Explain, using a diagram if required, why a regular pentagon will not tessellate. 🖩 [2 marks]

....

....

....

....

2 The internal angles of a regular polygon are 157.5°

How many sides has the polygon? .... [2 marks]

3 *ABIJKL* and *BCDEFGHI* are regular polygons. Calculate angle *JIH*. [3 marks]

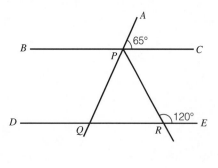

....

4 Calculate angle *FED*. State all your reasons. 🖉 [3 marks]

....

....

....

....

5 Lines *BC* and *DE* are parallel. Calculate angle *CPR*, giving all your reasons. 🖉 [2 marks]

....

....

....

Score /12

For more help on this topic, see Letts GCSE Maths Higher Revision Guide pages 58–59.

1 Draw lines to match each triangle with its correct description. [3 marks]

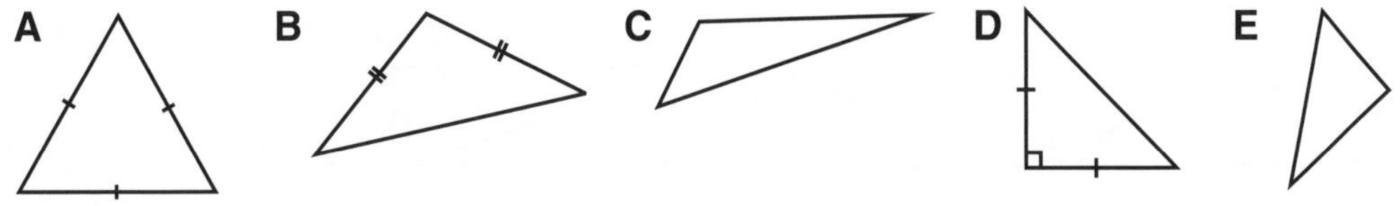

A **B** **C** **D** **E**

| Scalene | Right-angled isosceles | Equilateral | Obtuse scalene | Obtuse isosceles |

2 Use this diagram to prove the angle sum of a triangle is 180°. [3 marks]

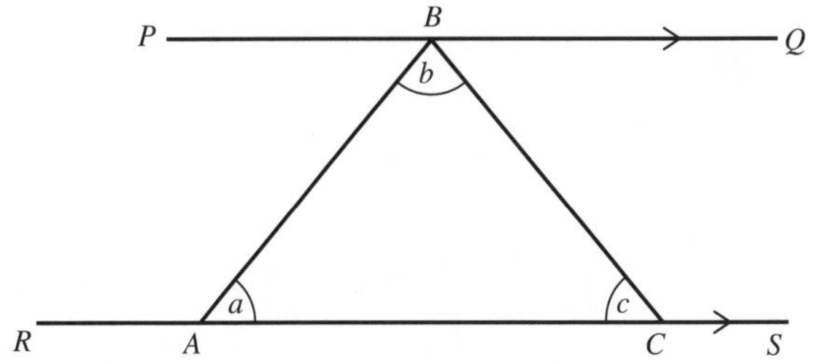

P *B* *Q*
b
a *c*
R *A* *C* *S*

....

....

....

....

3 Use this diagram to find the angle sum of an octagon. [3 marks]

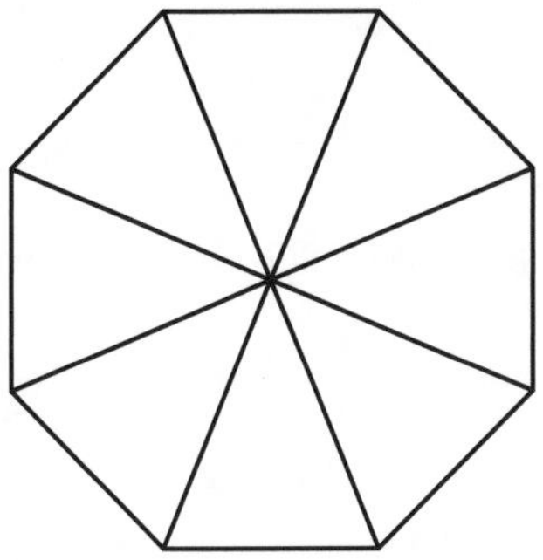

....

4 Complete this sentence with the name of the correct quadrilateral.

A .... has two pairs of equal sides but only one pair of equal angles. [1 mark]

5 Find angle *x*, giving all your reasons. [3 marks]

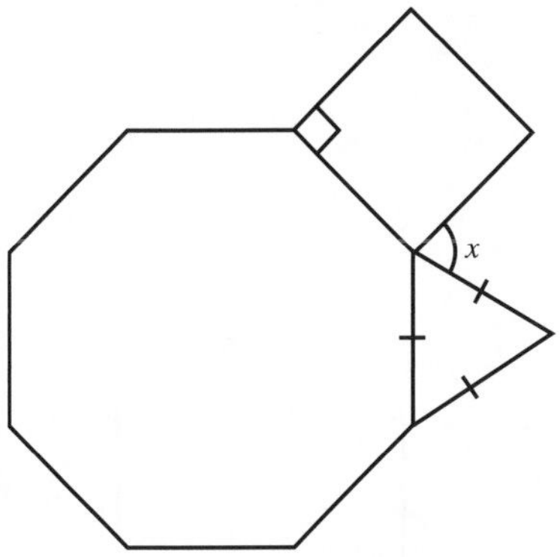

x

....

....

....

....

Score /13

For more help on this topic, see Letts GCSE Maths Higher Revision Guide pages 60–61.

Module 26

1 Circle the two similar triangles. [1 mark]

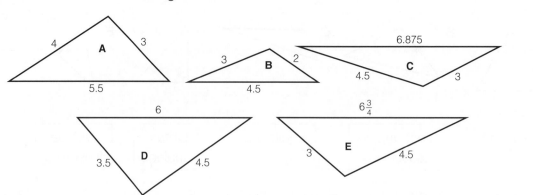

2 Circle the two congruent triangles. [1 mark]

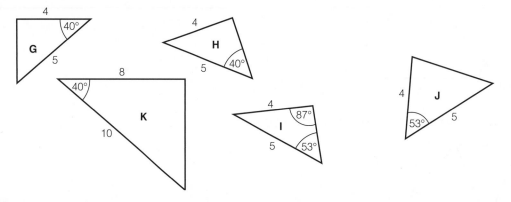

3 *PQRS* is a parallelogram. Prove that triangle *PQS* is congruent to triangle *RSQ*. [3 marks]

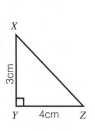

....

....

....

4 State whether these triangles are congruent and give your reasons. [2 marks]

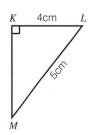

....

....

....

5 Cylinder A and cylinder B are mathematically similar.

The volume of cylinder A is 320cm³. What is the volume of B? [4 marks]

.... cm³

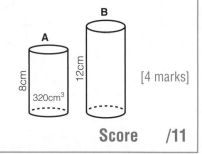

Score **/11**

For more help on this topic, see Letts GCSE Maths Higher Revision Guide pages 62–63.

1 (a) Describe the transformation
that moves A on to B. [2 marks]

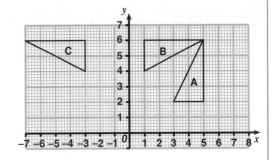

....

....

....

(b) Describe the transformation that moves B on to C. [2 marks]

....

2 Describe fully the single transformation
that maps P on to Q. [3 marks]

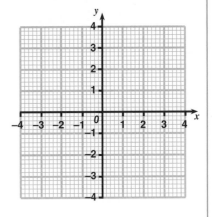

....

....

....

3 (a) Plot the triangle T with coordinates
(3, 3), (3, 1) and (4, 1). [1 mark]

(b) Rotate T 90° clockwise with centre
(0, 0) and label the image V. [2 marks]

(c) Reflect V in the *y*-axis and label
the image W. [2 marks]

(d) What single transformation maps T
directly on to W? [2 marks]

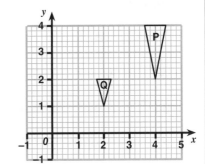

....

4 Enlarge this shape with

(a) scale factor 2, centre (3, 5) [2 marks]

(b) scale factor 0.5, centre (0, 0) [2 marks]

(c) scale factor –1, centre (0, 2). [2 marks]

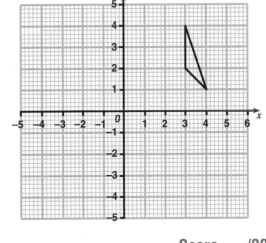

Score   /20

For more help on this topic, see **Letts GCSE Maths Higher Revision Guide pages 64–65.**

Transformations

Module 28

1 Complete the following statements with the correct word.

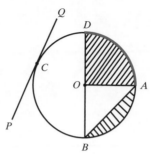

(a) *OA* is the .... of the circle.   [1 mark]

(b) *BD* is a ....   [1 mark]

(c) *PCQ* is a .... to the circle at *C*.   [1 mark]

(d) The curved line around the full circle is the ....   [1 mark]

(e) The blue line *AD* is an ....   [1 mark]

(f) The shaded area *OAD* is a ....   [1 mark]

(g) The shaded area between *A* and *B* is a ....   [1 mark]

(h) *OAB* is an .... triangle.   [1 mark]

2 Explain, with reasons, whether each statement is **true** or **false**.

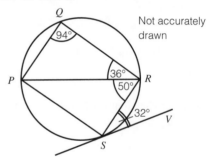

Not accurately drawn

(a) *PR* is a diameter of the circle.   [1 mark]

....

....

(b) Angle *PSR* = 86°   [1 mark]

....

....

(c) *SV* is a tangent to the circle.   [2 marks]

....

....

3 *PQRS* are points on a circle centre *O*.

Work out the size of these angles, giving reasons for your answers.

(a) *PRS* [2 marks]

....

....

....

(b) *PRQ* [3 marks]

....

....

....

....

4 **(a)** Calculate these angles, giving reasons for your answers.

(i) *ABD* [2 marks]

....

....

....

....

(ii) *BDA* [2 marks]

....

....

(b) Prove that triangle *ABD* is isosceles. [2 marks]

....

....

(c) Prove that triangles *CDF* and *ABF* are similar. [2 marks]

....

....

....

Score /25

For more help on this topic, see Letts GCSE Maths Higher Revision Guide pages 66–67.

Circles

Module 29

1 What 3D shape has 6 faces, 10 edges and 6 vertices? [2 marks]

....

2 Sketch two different 3D solids with five faces and name them. [3 marks]

3 Draw the plan and elevations of this 3D shape. [4 marks]

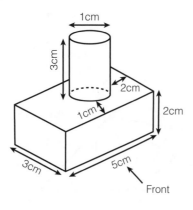

1cm

3cm

2cm

1cm

2cm

3cm

5cm

Front

4 Sketch the 3D shape shown by this plan and elevations. [3 marks]

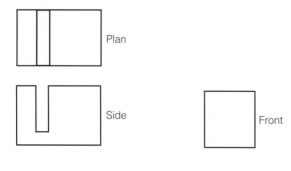

Plan

Side

Front

Score /12

For more help on this topic, see Letts GCSE Maths Higher Revision Guide pages 68–69.

1 Calculate the area and perimeter of these shapes.

(a) [3 marks]

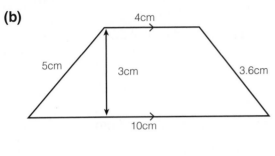

Area: .... m²   Perimeter: .... m

(b) [3 marks]

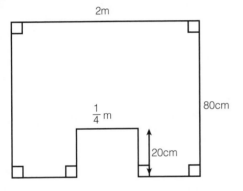

Area: .... cm²   Perimeter: .... cm

2 Calculate the area of this shape. [2 marks]

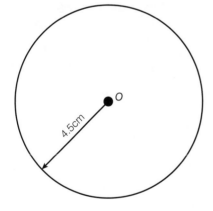

....

3 Calculate the circumference of this circle. Give your answer to 2 significant figures. [3 marks]

.... cm

4 Calculate the area of a circle with diameter 8cm. Give your answer in terms of π. [3 marks]

....

5 What is the perimeter of this shape?   [2 marks]

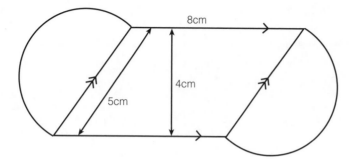

.... cm

6 Calculate the surface area of a cylinder with radius 4cm and height 5cm.
Give your answer in terms of π. 🖩   [3 marks]

.... cm²

7 This shape is a cone of vertical height 4cm sitting on a cube of side 3cm.
Calculate its volume.   [3 marks]

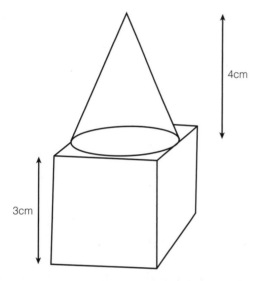

....

8 A sphere of radius 3.5cm and a cube with sides a cm have the same volume.
Find a. Give your answer to 3 significant figures.   [3 marks]

.... cm

9 Calculate the volume of a hemisphere with radius 5cm.
Give your answer in terms of π.   [3 marks]

....

Score   /28

For more help on this topic, see Letts GCSE Maths Higher Revision Guide pages 70–71.

1 (a) Work out the size of angle p. [3 marks]

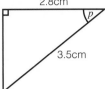

2.8cm

p

3.5cm

....

(b) Find the length KL. Give your answer to 3 significant figures. [3 marks]

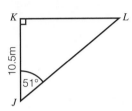

K L

10.5m

51°

J

.... m

2 Calculate angle ABC. [3 marks]

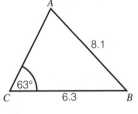

A

8.1

63°

C 6.3 B

....

3 Find q, leaving your answer as a square root. 📱 [3 marks]

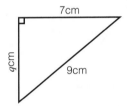

7cm

$q$cm

9cm

....

4 (a) Write down the value of 📱

   (i) sin 45° .... [1 mark]

   (ii) tan 45° .... [1 mark]

(b) Calculate a and b. 📱 [4 marks]

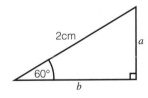

2cm

a

60°

b

$a =$ .... $b =$ ....

5 Find the length of the largest pencil that can just fit in a box 14cm by 6cm by 3cm. [2 marks]

.... cm

Score /20

For more help on this topic, see Letts GCSE Maths Higher Revision Guide pages 72–73.

1 Here are base vectors **r**, **s** and **t** drawn on isometric paper.

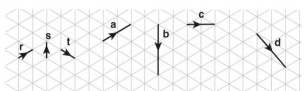

Write each of the vectors **a** to **d** in terms of **r**, **s** and **t**. [4 marks]

a = ...   **b** = ...   **c** = ...   **d** = ...

2 *ABCD* is a parallelogram.
BE is a straight line with *AB = AE*.
$\overrightarrow{AB}$ = **b**, $\overrightarrow{AD}$ = **d**.

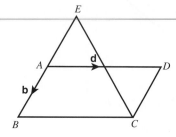

(a) Write $\overrightarrow{BE}$ in terms of **b** and **d**.   .... [1 mark]

(b) Write $\overrightarrow{AC}$ in terms of **b** and **d**.   .... [2 marks]

(c) Write $\overrightarrow{CE}$ in terms of **b** and **d**.   .... [2 marks]

3 *PQR* is an equilateral triangle.
OPQ is an isosceles triangle with *OQ = QR*

$\overrightarrow{OP}$ = **p**

$\overrightarrow{OQ}$ = **q**

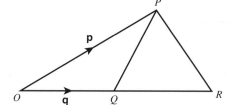

Find $\overrightarrow{OR}$, $\overrightarrow{PQ}$ and $\overrightarrow{PR}$. [3 marks]

$\overrightarrow{OR}$ = ... $\overrightarrow{PQ}$ = ... $\overrightarrow{PR}$ = ...

4 If *A* is the midpoint of *OP* and *B* is the midpoint of *OQ*, use vectors to
prove that *AB* is parallel to *PQ*. [3 marks]

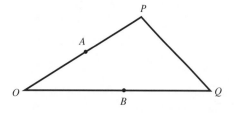

....

....

Score   /15

For more help on this topic, see Letts GCSE Maths Higher Revision Guide pages 74–75.

1 There are three bananas, five oranges, four apples and two mangoes in a fruit bowl. A piece of fruit is taken at random.

Find the probability that it is

(a) an apple .... [1 mark]

(b) a banana or a mango .... [1 mark]

(c) not an orange. .... [1 mark]

2 Paige wants to test whether she can predict the suit of a card drawn from an ordinary shuffled deck of playing cards. She does 120 tests.

How many correct predictions do you think she will make? [2 marks]

....

3 The table shows the probability that Mark will choose a certain activity when he gets home from school on any given day.

Activity	Watch TV	Do homework	Play on computer	Go for a bike ride	Other
Probability	0.2		0.3	0.25	0.1

Find the probability that Mark will do homework on any given day. [2 marks]

....

4 Brian bought a dice from the joke shop. He rolled it 25 times and recorded the results. Here are the results:

Number	1	2	3	4	5	6
Frequency	0	3	2	4	4	12

(a) Find the relative frequency of rolling a 6. [1 mark]

....

Brian rolls the dice another 250 times.

(b) How many times would he be expected to get an even number? [3 marks]

....

5 Morgan spins the spinner until she gets a 4.

Find the probability that she will stop after exactly two spins. [2 marks]

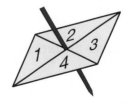

....

Score /13

For more help on this topic, see Letts GCSE Maths Higher Revision Guide pages 78–79.

1 Fifty students are voting for a new class president. They can each vote for a maximum of two people from Luke, Dan and Jack.

Luke gets 26 votes altogether
7 students vote for Dan and Luke
2 students vote for Dan and Jack

Dan gets 17 votes altogether
4 students vote for Luke and Jack
3 students don't vote

(a) Complete the Venn diagram to show this information. The universal set ξ contains all 50 students.

[3 marks]

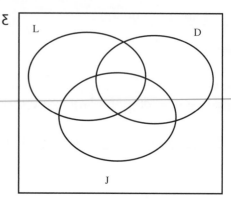

(b) Find the probability that a student chosen at random votes for Jack only. [1 mark]

....

(c) Find the probability that a student who voted for Luke also voted for Dan. [2 marks]

....

2 Katy is a midwife. If she works on Christmas Day, then the probability that she will have to work on New Year's Day is 0.25

If she does not work on Christmas Day, then the probability that she will have to work on New Year's Day is 0.85

The probability that Katy will have to work on Christmas Day is 0.7

(a) Complete the tree diagram. [2 marks]

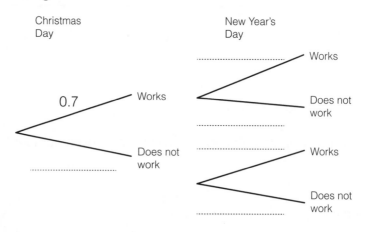

(b) Calculate the probability that Katy will only have to work on one of the two days. [2 marks]

....

Score /10

For more help on this topic, see Letts GCSE Maths Higher Revision Guide pages 80–81.

1 A washing powder manufacturer conducts a survey of the actual weight of powder in the 500g boxes on the production line. A sample of 12 boxes is taken over a period of 1 hour:

495, 503, 490, 505, 490, 498, 500, 510, 505, 498, 498, 501

(a) What is the mode for the sample? [1 mark]

....

(b) Find the median. [1 mark]

....

(c) The manufacturer produces 10 000 boxes per hour.
Give two reasons why this may not be a good sample. [2 marks]

....

....

2 The table shows the weights of 160 different species of birds in the forest.

Weight (g)	Frequency
$30 \leqslant w < 40$	20
$40 \leqslant w < 50$	25
$50 \leqslant w < 60$	22
$60 \leqslant w < 80$	44
$80 \leqslant w < 100$	32
$100 \leqslant w < 120$	15
$120 \leqslant w < 150$	2
	160

(a) Which class interval contains the median? [1 mark]

....

(b) Calculate an estimate of the mean weight of the birds. [3 marks]

....

(c) What percentage of the birds weigh less than 80g? [2 marks]

....

3 There are five cards with an integer on each. The mean of all the numbers on the cards is 7. Both the mode and the median of the numbers are 6.

What numbers could be on the cards? 📝 [3 marks]

....

Score   /13

For more help on this topic, see Letts GCSE Maths Higher Revision Guide pages 84–85.

1 The table shows the sales figures for a car dealership over two different weeks.

Sales	Week 1	Week 2
Monday	10	25
Tuesday	25	20
Wednesday	29	15
Thursday	40	19
Friday	16	

(a) There were 100 car sales in total in week 2. Complete the table. [1 mark]

(b) Show this data in a suitable chart. [3 marks]

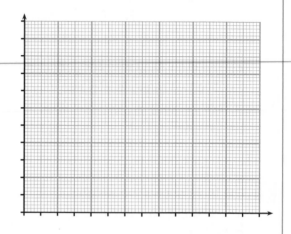

2 The manager of a pizza shop conducts a survey of favourite pizzas. She wants to show this information on a pie chart.

	Frequency	Angle
Vegetarian	45	60°
Seafood	80	
Meat	50	
Chicken	40	
Mushroom		

Complete the table. [2 marks]

3 The table shows the sales of a particular type of mobile phone over an eight-year period.

Year	1	2	3	4	5	6	7	8
Sales (millions)	1.4	11.6	20.7	40	72.3	125	150.3	169.2

(a) By plotting a suitable graph, use this information to predict the sales for year 9.

.... [4 marks]

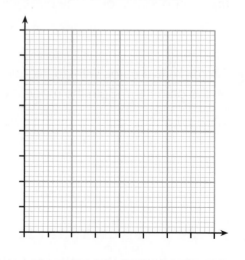

(b) Comment on the reliability of using your graph to predict sales in year 14. [2 marks]

....

....

4 The speeds of 12 drivers and their ages were recorded in the table below. John says that 'younger drivers drive too fast'.

Age (years)	20	32	24	30	22	40	35	34	42	22	38	32
Speed (mph)	38	30	37	32	39	30	32	33	28	36	27	34

(a) Use an appropriate diagram to comment on whether the data supports John's conclusion.

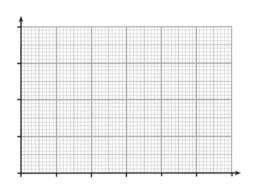

.... [3 marks]

(b) Describe the relationship between age and driving speed.

.... [1 mark]

5 Wasim hypothesises that students travel further to get to college than they used to. In a survey in 2005, the mean weekly travelling distance to a college was 25.8km. The histogram below shows the results of another survey taken in 2015.

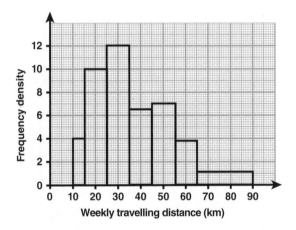

Weekly travelling distance

Does the evidence support the hypothesis? [5 marks]

....

....

....

Score /21

For more help on this topic, see Letts GCSE Maths Higher Revision Guide pages 86–87.

Module 37

1 The test scores of 24 students in a history exam are given below.

Class A	33	45	67	83	56	23	57	45	73	43	26	35
Class B	33	27	40	44	78	28	49	38	54	32	51	74

(a) Calculate the mean for each of the two classes. [2 marks]

....

....

(b) Draw box plots for both classes. [4 marks]

....

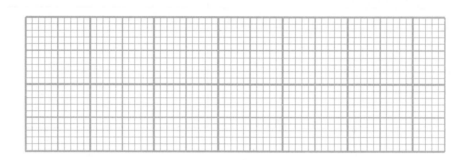

(c) Which class did better? Give reasons for your answer. [3 marks]

....

....

....

2 The table below shows the results of a survey on the cost of car insurance for 30–50-year-olds.

(a) Construct a cumulative frequency curve to represent the data. [3 marks]

Cost (£)	Frequency
$100 \leqslant c < 150$	10
$150 \leqslant c < 200$	38
$200 \leqslant c < 250$	48
$250 \leqslant c < 300$	31
$300 \leqslant c < 350$	20
$350 \leqslant c < 400$	8
$400 \leqslant c < 500$	5

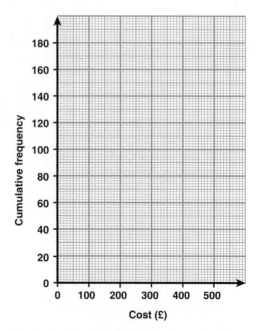

Comparing Distributions

Module 38

(b) Draw a box plot for this data. [2 marks]

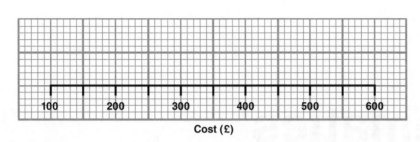

Cost (£)

(c) The same survey was conducted for drivers younger than 30 years old. The results are shown on this plot.

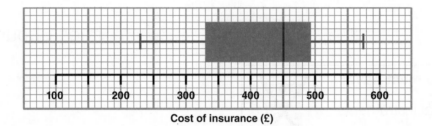

Cost of insurance (£)

Make two comparisons between the two sets of data. [2 marks]

Comparison 1: ....

....

Comparison 2: ....

....

(d) Use your cumulative frequency curve to estimate how many 30–50-year-olds pay more than £325 for car insurance. Explain why this is an estimate. [2 marks]

....

....

....

....

Score /18

For more help on this topic, see Letts GCSE Maths Higher Revision Guide pages 88–89.

Module 38

GCSE
Mathematics
Higher tier

Paper 1   Time: 1 hour 30 minutes

For this paper you must have:

- mathematical instruments

You must **not** use a calculator.

Instructions

- Use black ink or black ball-point pen. Draw diagrams in pencil.
- Read each question carefully before you start to write your answer.
- Diagrams are not accurately drawn unless otherwise stated.
- Answer **all** the questions.
- Answer the questions in the space provided.
- In all calculations, show clearly how you work out your answer. Use a separate sheet of paper if needed. Marks may be given for a correct method even if the answer is wrong.

Information

- The mark for each question is shown in brackets.
- The maximum mark for this paper is 80.

Name: _____

1. Which average is affected most by an outlier? [1]

....

2. **(a)** Dorcas is thinking of a number. She multiplies it by 4 and then adds 3. She gets the answer −5.

What number was she thinking of? [2]

....

(b) Ketsia starts with the number 3. She takes away a and then multiplies by b.

She gets an answer of 28. a is a negative integer and b is a positive integer.

Find two possible pairs of values for a and b. [2]

$a =$ .... $b =$ ....

$a =$ .... $b =$ ....

3. George, Timmy and Ann shared some money in the ratio 5 : 7 : 9

Ann got £32 more than Timmy.

How much money did George get? [3]

£ ....

4. Jenni plays violin in an orchestra. She has 22 performances this year. For each performance she needs to get a return train ticket which costs £8.85. The orchestra pays $\frac{1}{4}$ of her travel costs.

(a) Estimate how much Jenni will have to pay in travel costs for the year. [2]

£ ....

(b) Calculate the exact cost of the train tickets for the year. [2]

£ ....

5. Solve the simultaneous equations $8x - 5y = 19$ [4]
   $12x + y = 3$

$x =$ .... $y =$ ....

6. There are some orange and mint chocolates in a bag. The probability of taking a mint chocolate is $\frac{1}{3}$

Amorreane takes a mint at random and eats it.

The probability of taking a mint chocolate is now $\frac{1}{4}$

How many orange chocolates are in the bag? [2]

....

7. Increase £340 by 15% [2]

....

8.

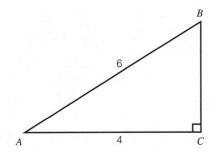

(a) Write down the value of cos A. [1]

....

(b) What is the value of tan B? Leave your answer as a square root in its simplest form. [4]

....

9. Work out $\dfrac{5}{8} \div \dfrac{2}{3}$ [1]

....

10. Solve the equation $(x+1)(x-9)=11$ [4]

....

11. Simplify $\left(3x^2 y^5\right)^4$ [2]

....

12. The total TV sales over a five-year period for a national electrical store were 120 000 units. Find the sales figures for computers and TVs in year 4. **[3]**

Year	Computer sales (thousands)	TV sales (thousands)
1	12	
2	14	
3	13	
4	a	$2a+5$
5	16	

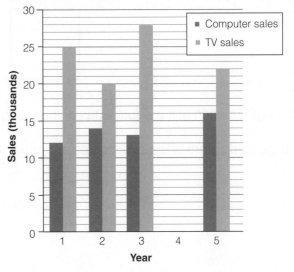

TV sales: ....

Computer sales: ....

13. Find the equation of the line that is parallel to $y = 2x + 7$ and passes through the point (0, –3). **[2]**

....

14. Here is a square and an isosceles triangle.

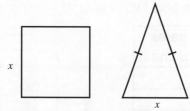

The length of each of the equal sides of the triangle is 3cm greater than the side of the square.

(a) If the perimeters of the two shapes are equal, find the value of x. **[3]**

....

(b) Show that the height of the triangle is equal to the diagonal of the square. **[3]**

....

....

....

....

15. The graph of $y = \cos x$ is shown for $0° \leqslant x \leqslant 360°$

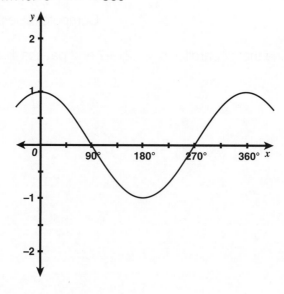

On the same grid, sketch and label the graphs of

(a) $y = -\cos x$ **[2]**

(b) $y = \cos x + 1$ **[2]**

16. Calculate the circumference of this circle.

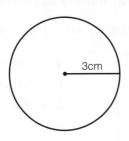

Leave your answer in terms of π. [2]

.... cm

17. Calculate angle *BCD*, giving your reasons. [3]

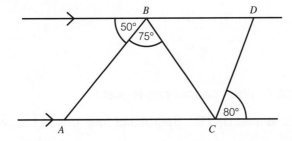

....

....

....

....

18. Work out the next term of this quadratic sequence:

    −2   3   14   31 [2]

....

19. Mike has 240 different films in his collection. He has some films on both Blu-ray and DVD.

The Venn diagram shows the information.

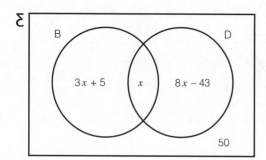

ℰ = 240 films in the collection
B = films on Blu-ray
D = films on DVD

Mike watches one of his films.

Find the probability that he chooses a film which he has on both Blu-ray and DVD. **[4]**

....

20. Stereo headphones A cost £a and headphones B cost £b.

When a and b are both increased by £20, the ratio of their prices becomes 5 : 2

When a and b are both decreased by £5, the ratio of their prices becomes 5 : 1

Find the ratio $a : b$ in its lowest terms. **[4]**

....

21. (a) Write $\sqrt{48}$ as a surd in its simplest form. [1]

....

(b) Write $\dfrac{2}{3 + \sqrt{7}}$ in the form $a + \sqrt{b}$ [2]

....

22. Which is more, 15% of 260 or 18% of 210?

Show your working. [3]

....

....

....

....

23. The results of a survey on the lifetime of two different light bulbs are shown on the box plots. [2]

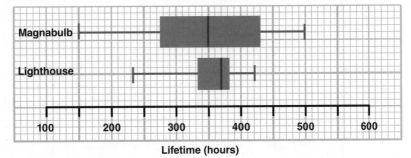

Lifetime (hours)

Give two reasons why Lighthouse bulbs might be better.

Reason 1: ....

....

Reason 2: ....

....

24. Two cones A and B have volumes of 240cm³ and 810cm³.

If the surface area of cone A is 180cm², what is the surface area of cone B? **[4]**

.... cm²

25. Sketch the graph of $y = x^2 - 8x + 17$, showing clearly the coordinates of the turning point and the coordinates of any intercepts with the coordinate axes.

Write down the equation of the line of symmetry. **[6]**

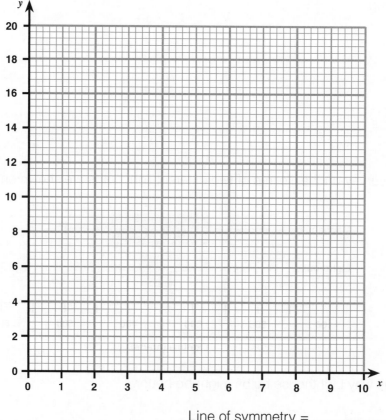

Line of symmetry = ...

GCSE
Mathematics
Higher tier

Paper 2

Time: 1 hour 30 minutes

For this paper you must have:

- a calculator
- mathematical instruments

Instructions

- Use black ink or black ball-point pen. Draw diagrams in pencil.
- Read each question carefully before you start to write your answer.
- Diagrams are not accurately drawn unless otherwise stated.
- Answer **all** the questions.
- Answer the questions in the space provided.
- In all calculations, show clearly how you work out your answer. Use a separate sheet of paper if needed. Marks may be given for a correct method even if the answer is wrong.
- If your calculator does not have a π button, take the value of π to be 3.142 unless the question instructs otherwise.

Information

- The mark for each question is shown in brackets.
- The maximum mark for this paper is 80.

Name: ....

1. A new tablet computer is released and the older model costing £425 is reduced by 12%.

    What is its new price? [2]

    £ ....

2. Find two numbers, **greater than 71**, which have the following **two** properties:

    - a multiple of 5 and 7
    - a common factor of 420 and 630 [4]

    ....

3. Calculate the area of this shape. Give your answer to 3 significant figures. [3]

    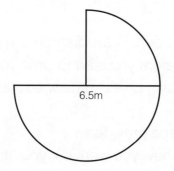

    6.5m

    ....

4. A biologist conducted a survey of different lakes to investigate algae growth and the amounts of nitrate present. She recorded the concentrations of both algae and nitrate in a table.

Algae (cells/ml)	10	12	55	20	5	24	42	44	15	30	22	38
Nitrate (mg/L)	0.5	1.5	3.6	0.7	0.5	1.8	1.5	1.8	1.7	2.3	0.8	2.6

(a) Draw a scatter graph of this data on the grid below. **[2]**

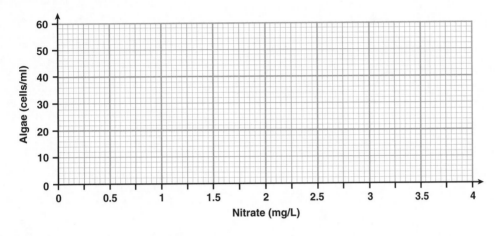

(b) What type of correlation is shown between nitrate levels and algae concentrations? **[1]**

....

(c) Use the graph to estimate the nitrate levels in a lake with an algae concentration of 28 cells/ml. **[2]**

.... mg/L

(d) The biologist wants to use this data to predict algae concentrations for any lake.

Comment on the biologist's idea. **[1]**

....

....

5. Expand

   (a) $(2x-1)^2$ [2]

   (b) $(2x-1)^3$ [2]

6. **(a)** Work out the following. Write down all the figures on your calculator display. [2]

   $$\frac{\sqrt{35}+6^2}{(2-0.04)^3}$$

   (b) Round your answer in part (a) to 3 significant figures. [1]

7. Show that the point $(-3, -2)$ is on the curve $y = x^2 + 2x - 5$ [2]

8. Three feeding bowls, A, B and C, are placed in a cage with two hamsters. The hamsters are equally likely to choose any of the feeding bowls.

Find the probability that both hamsters choose bowl C. **[2]**

....

9. Matthew is estimating the height of his house using the Sun.

A stick 1m long casts a shadow 60cm long. At the same time the shadow of the house is 5.4m long.

How tall is the house? **[3]**

....

10. Sketch the graph of $y = 2^{-x}$ on the axes below. [2]

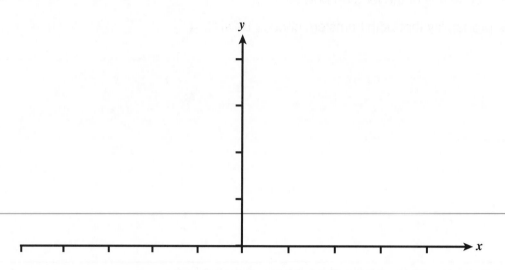

11. The graph shows the temperature of a jug of custard as it cools.

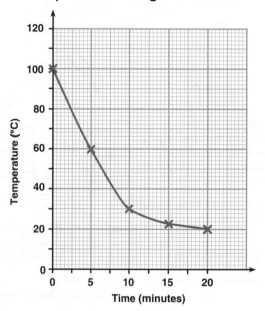

Temperature change of custard

(a) What is the approximate temperature of the room? [1]

....

(b) What is the temperature of the custard after 15 minutes? [1]

....

(c) What is the rate of cooling after 10 minutes? [3]

....

12. Jerome is a salesman. He visits farms to try to sell gates and fencing equipment.

When Jerome visits a farm, the probability that he will make a sale is 0.4

One morning Jerome visits two farms.

(a) Complete the tree diagram to show all the outcomes. [2]

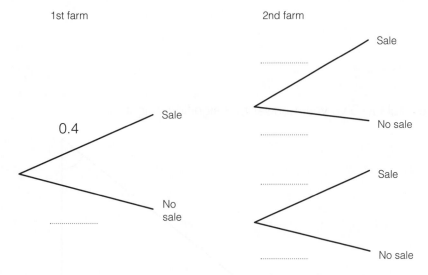

1st farm   2nd farm

0.4   Sale

No sale

Sale

No sale

Sale

No sale

(b) Find the probability that Jerome sells equipment to at least one of the farms. [2]

13. Rearrange $y = \dfrac{10x - 2}{5 - x}$ to make x the subject.   **[4]**

....

14. Calculate the angle R. Give your answer to 3 significant figures.   **[3]**

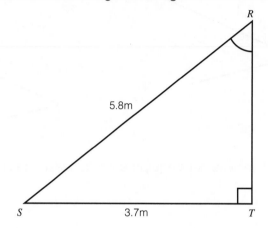

....

15. Copper and zinc are mixed to form an alloy in the ratio 9 : 7 by mass.

(a) If 27kg of copper are used to make a batch of alloy, how much alloy can be made? **[3]**

.... kg

(b) The density of copper is 8900kg/m³.

Find its density in g/cm³. **[4]**

.... g/cm³

(c) If the density of zinc is 7g/cm³, work out the density of the alloy. **[3]**

.... g/cm³

16. Find the coordinates of the points where the line $y = 2x - 4$ crosses the curve $y = x^2 - 7x + 14$ [4]

17. Prove that $0.08\dot{1} = \dfrac{9}{110}$ [3]

18. Work out the equation of the line that is perpendicular to the line $y = 7 - 2x$ and passes through the point (8, –1). [4]

19. The ratio of the radii of two cones is 1 : 3

    (a) Calculate the ratio of the curved surface areas of the cones.   **[1]**

    ....

    (b) If the volume of the larger cone is 10.8 litres, what is the volume of the smaller cone?   **[2]**

    ....

20. Find the length *KL*.   **[3]**

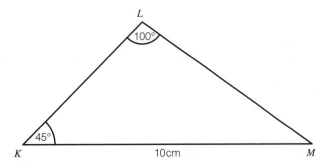

....

21. The nth term in an arithmetic sequence is given by $U_n = 5n + 2$

Another sequence is given by $V_n = 4U_{n+2} - 2U_n$

Prove that V_n is also an arithmetic sequence.

Is there any value for n such that $U_n = V_n$?   [6]

....

....

....

....

....

....

GCSE
Mathematics
Higher tier

Paper 3

Time: 1 hour 30 minutes

For this paper you must have:
- a calculator
- mathematical instruments

Instructions

- Use black ink or black ball-point pen. Draw diagrams in pencil.
- Read each question carefully before you start to write your answer.
- Diagrams are not accurately drawn unless otherwise stated.
- Answer **all** the questions.
- Answer the questions in the space provided.
- In all calculations, show clearly how you work out your answer. Use a separate sheet of paper if needed. Marks may be given for a correct method even if the answer is wrong.
- If your calculator does not have a π button, take the value of π to be 3.142 unless the question instructs otherwise.

Information

- The mark for each question is shown in brackets.
- The maximum mark for this paper is 80.

Name: ....

1. Lyse and Lysette each test out a biased coin to find out the estimated probability of it landing on heads. Here are the results:

	Number of coin flips	Number of heads
Lyse	40	16
Lysette	80	28

(a) Whose results give the best estimate for the probability of getting heads?

Explain your answer. [1]

....

....

....

....

The coin is flipped 500 times.

(b) How many times do you expect the coin to land on heads? [2]

....

2. Find an integer value of x satisfying $3x + 4 > 19$ and $2x - 1 < 13$ [3]

....

3. Sketch two graphs to show

(a) y directly proportional to x. **[2]**

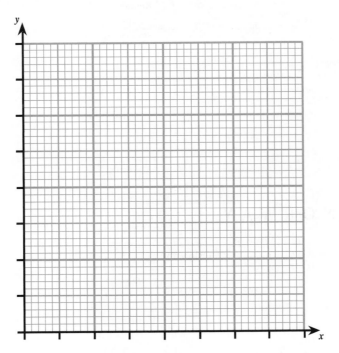

(b) y inversely proportional to x. **[2]**

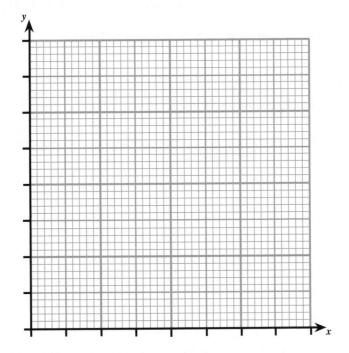

4. Here is a map of part of Norway.

Scale 1 : 6 500 000

(a) Estimate the straight-line distance from Trondheim to Bergen. [2]

....

(b) The actual road and ferry journey is 700km.

Express the road and ferry distance as a percentage of the straight-line distance. [2]

....

(c) If it takes 11 hours to make the journey by car, what is the average speed? [3]

....

5. **(a)** Write $3^5 \div 3^{10}$ as a power of 3. [1]

....

(b) Write $0.000\,070\,2$ in standard form. [1]

....

(c) Evaluate $16^{\frac{1}{2}}$ [1]

....

6. A farmer measures the growth (measured as a change in height) of his crop over a period of eight weeks. The results are shown in the table below.

Growth (cm)	Frequency
$0 \leqslant h < 20$	11
$20 \leqslant h < 40$	38
$40 \leqslant h < 60$	49
$60 \leqslant h < 80$	30
$80 \leqslant h < 120$	12

(a) Which is the modal group? [1]

....

(b) Find the estimated mean growth for the crop over this period. [3]

.... cm

(c) The farmer says that over one-third of his crop grew by more than 60cm during this period.

Does the data support the farmer's claim? [2]

....

....

....

....

....

....

7. The International Space Station (ISS) travels at a speed of 4.48×10^4 km/h.

(a) Write this value as an ordinary number. [1]

....

It takes three hours for the ISS to orbit the Earth twice.

(b) Find the distance that the ISS travels in one full orbit around the Earth.

Write your answer in standard form. [3]

....

8.

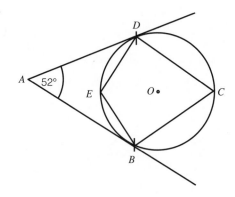

(a) Find angle *BCD*. [3]

....

(b) Find angle *BED*. [2]

....

9. The value of a motor bike (£V) is given by the formula $V = 18\,000 \times 0.85^t$ where t is the age in whole number of years.

   (a) What is the value of the bike when it is new? [1]

   ....

   (b) What is the bike worth after two years? [2]

   ....

   (c) After how many years will its value be below £10 000? [2]

   ....

10. Solve the equation $\dfrac{6}{x} - \dfrac{4}{2x+2} = 2$, giving your answers to 2 decimal places. [5]

   ....

11. A hot-air balloon flies from home (H) on a bearing of 080° for 200km to a point D.

At D it turns on to a bearing of 200° for a further 300km to a point E.

Calculate the distance and bearing which will return the balloon home. **[6]**

12. The volume of a cone is given by $\frac{1}{3}\pi r^2 h$

A frustum is the shape left when the top of a cone has been cut off.

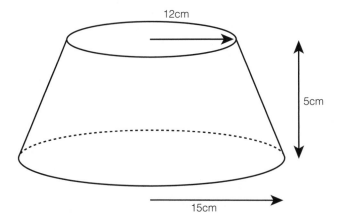

12cm

5cm

15cm

Find the volume of this frustum. Give your answer to 3 significant figures. **[5]**

13. Write $4x^2 - 24x + 41$ in the form $a(x-b)^2 + c$ **[4]**

14. Robson is taking part in a mountain bike race. He must finish in the top 10 to qualify.

The probability that he will qualify if it is dry is 0.8

The probability that he will qualify if it is raining is 0.7

The probability that it will be raining on the day of the race is also 0.7

Find the probability that he will qualify in the race. **[3]**

15. A vicar has one full bottle of communion wine for a midnight mass service.

The bottle holds 750ml of wine to the nearest 5ml.

Each communion cup holds 5ml of wine to the nearest ml.

If he uses the full bottle, what is the smallest number of cups that the vicar will fill for the midnight mass service? **[3]**

....

16. Functions f and g are given by $f(x) = 3x - 1$ and $g(x) = 1 + 2x$

(a) Find $fg(3)$ **[2]**

....

(b) Solve the equation $f^{-1}(x)g^{-1}(x) = 4$ **[5]**

....

17. Evaluate $16^{\frac{3}{4}}$　　　　　　　　　　　　　　　　　　　　　　　　　　　**[2]**

18. An object travels in a straight line from rest such that its velocity v at time t is shown by the velocity–time graph below.

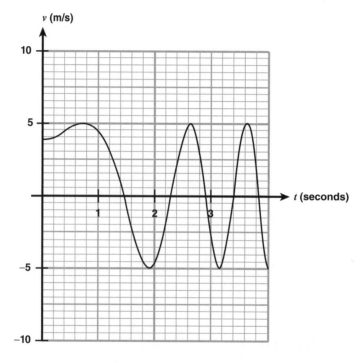

Estimate the acceleration of the object after one second.　　　　　　　　　**[5]**

.... m/s²

You are encouraged to show your working out, as you may be awarded marks for method even if your final answer is wrong. Full marks can be awarded where a correct answer is given without working being shown but if a question asks for working out, you must show it to gain full marks. If you use a correct method that is not shown in the mark scheme below, you would still gain full credit for it.

Module 1: Place Value and Ordering

1. (a) -7 (b) 21   (c) -3

> Remember the rules for signs and BIDMAS.

2. (a) False   (b) True   (c) True   (d) False
3. (a) $-3°C$ (b) $-9°C$
4. (a) 5.418   (b) 5418   (c) 12900   (d) 420

> Compare the position of decimal points with the original calculation.

5. (a) £35 [Accept –£35]
   (b) £35 – £14 [1] = £21 [1] [Accept –£21]

Module 2: Factors, Multiples and Primes

1. (a) 1 and 3   (b) 1, 4 and 16   (c) 6, 12, 24 and 48
2. 21 [1] or 45 [1]

> Test the prime numbers between 1 and 50.

3. (a) 55   (b) 7
4. (a) $72 = 2 \times 2 \times 18 = 2 \times 2 \times 2 \times 3 \times 3$ [2] [Two correct steps of decomposition get 1 mark] $= 2^3 \times 3^2$ [1]

> Use factor trees.

   (b) $90 = 2 \times 3 \times 15$ [1] $= 2 \times 3 \times 3 \times 5$ [1] $= 2 \times 3^2 \times 5$ [1]
   (c) $2 \times 3 \times 3$ [1] $= 18$ [1]

> Look for the common factors from the lists in (a) and (b).

   (d) $18 \times 2 \times 2 \times 5$ [1] $= 360$ [1]

> HCF × remaining factors.

5. Bus A: 12   24   36   48   60   72   84
   Bus B: 28   56   84   112   140   168   196 [1]
   LCM = 84 [1]
   8.00am + 84 minutes = 9.24am [1]

Module 3: Operations

1. (a) 53

> Multiplication before addition.

   (b) $\frac{11}{15}$

> Find a common denominator.

   (c) 4

2. $\frac{(-2)^2}{2}$ [1] $= 2$ [1]

3. (a) $\frac{12}{40}$ [1] $= \frac{3}{10}$ [1]   (b) $\frac{50}{18}$ [1] $= 2\frac{7}{9}$ [1]

> Find the reciprocal of the second fraction and multiply.

   (c) $1\frac{15-12}{20}$ [1] $= 1\frac{3}{20}$ [1]   (d) $\frac{3}{2} \times \frac{5}{19}$ [1] $= \frac{15}{38}$ [1]

> Convert to an improper fraction and divide as normal.

4. $1 - \left(\frac{2}{5} + \frac{3}{7}\right)$ [1] $= 1 - \frac{29}{35}$ [1] $= \frac{6}{35}$ [1]

5. $5\frac{1}{4} \times 3\frac{5}{6}$ [1] $= \frac{21}{4} \times \frac{23}{6}$ [1] $= \frac{483}{24} = 20\frac{1}{8}$m² [1]

6. $1 - \left(\frac{3}{5} + \frac{2}{9}\right)$ [1] $= 1 - \frac{37}{45}$ [1] $= \frac{8}{45}$ [1]

Module 4: Powers and Roots

1. $\sqrt[3]{64}$ $\sqrt[4]{625}$ 5^2 3^3 2^5
2. (a) $x = 7$ (b) $x = 125$ (c) $x = 3$

3. (a) 4^{10}

> Subtract the negative power, i.e. $4^{6-(-4)}$

   (b) $4\sqrt{2}$ (c) $\frac{1}{9}$

   $\sqrt{16} \times \sqrt{2} = 4\sqrt{2}$

4. (a) (i) 3.21×10^5 (ii) 6.05×10^{-4}
   (b) (i) 30×10^{-5} [1] $= 3 \times 10^{-4}$ [1]
   (ii) $41000 + 3400 = 44400$ [1] $= 4.44 \times 10^4$ [1]

5. (a) $\sqrt{3} \times 4 - \sqrt{3} \times \sqrt{3} = 4\sqrt{3} - 3$
   (b) $8 - 4\sqrt{5} + 12\sqrt{5} - 30$ [1] $= 8\sqrt{5} - 22$ [1]

6. (a) $\frac{4}{\sqrt{7}} \times \frac{\sqrt{7}}{\sqrt{7}} = \frac{4\sqrt{7}}{7}$

   (b) $\frac{5}{2-\sqrt{3}} \times \frac{2+\sqrt{3}}{2+\sqrt{3}}$ [1] $= \frac{5(2+\sqrt{3})}{(2-\sqrt{3})(2+\sqrt{3})} = \frac{10+5\sqrt{3}}{4+2\sqrt{3}-2\sqrt{3}-3}$ [1]

   $= \frac{10+5\sqrt{3}}{1} = 10 + 5\sqrt{3}$ [1]

Module 5: Fractions, Decimals and Percentages

1. $\frac{17}{51}$ $\frac{3}{8}$ 0.475 $\frac{12}{25}$

> Convert all the values to decimals to compare them.

2. $\frac{19}{40}$

3. $5 \div 8 = 0.625$ [1] $7 \div 11 = 0.\dot{6}\dot{3}$ [1]

   $\frac{7}{11}$ is closer to $\frac{2}{3}$ (as $\frac{2}{3} = 0.\dot{6}$) [1]

4. (a) $37 \times 49 = 1813$ [1]
   So $3.7 \times 4.9 = 18.13$ [1]
   (b) $3696 \div 14$ [1] $= 264$
   So $369.6 \div 1.4 = 264$ [1]

5. 34×129 [1] $= 4386$ [1]
   So total cost is £43.86 [1]

> Use long multiplication.

6. $\frac{4}{9}$

7. $99x = 45$ [1] $x = \frac{45}{99} = \frac{5}{11}$ [1]

8. $990x = 531$ [1] $x = \frac{531}{990} = \frac{59}{110}$ [1]

Module 6: Approximations

1. (a) 9.49   (b) 6.554   (c) 5.60

> Use a calculator first.

2. (a) 400000   (b) 0.0497   (c) 3.142

> Use calculator to find π.

3. $500 \div 35$ [1] $= 14.3$g [1]
4. $70 \times 40 \times 60$ [1] $= 168000$ [1]
   £1680 [1]
5. Upper bound = 1.735m [1]
   Lower bound = 1.725m [1]

> Half a centimetre above and below 1.73

6. 500.5×8 [1] $= 4004$ml [1]

> Use the upper bound to find the maximum amount.

7. Minimum value $= \frac{6.75}{0.345}$ [1] $= 19.57$ [1]

   Maximum value $= \frac{6.85}{0.335}$ [1] $= 20.45$ [1]

Module 7: Answers Using Technology

1. **(a)** 410.0625   **(b)** 7   **(c)** 5

2. $\dfrac{19.21}{32.76}$ **[1]** $= 0.586385836$ **[1]**

   Do not round at any point.

3. 1.2×10^{-9}

4. **(a)** $2\dfrac{20}{21}$

   Your calculator might show   `2⌐20⌐21`

   (b) $31\dfrac{1}{18}$

5. **(a)** $1.496 \times 10^8 \div 400$ **[1]** $= 374000$ **[1]** $= 3.74 \times 10^5$ km **[1]**
   (b) $1.392 \times 10^6 \div 3.48 \times 10^3$ **[1]** $= 400$ **[1]** $= 4 \times 10^2$ **[1]**
   (c) The Sun is 400 times bigger and 400 times further away (this is why a total eclipse appears the way it does).
   (d) $1.496 \times 10^8 + 3.74 \times 10^5$ **[1]** $= 149974000$ **[1]**
   $= 1.5 \times 10^8$ km **[1]**

Module 8: Algebraic Notation

1. **(a)** $5a + 2$ **(b)** $15h + 2k$ **(c)** $-2a + 2b$ **(d)** $8x^2 - x + 1$

2. **(a)** q **(b)** $27p^{12}$ **(c)** $6f^5g$ **(d)** $\dfrac{1}{8p^6}$

3. **(a)** $12k^2$ **(b)** a^3 **(c)** $p^{\frac{5}{2}}$

   (d) $p^{\frac{13}{2}}$ or $\sqrt{p^{13}}$

4. **(a)** $\dfrac{1}{64}$

   Remember to deal with the negative index first by using 'one over...'. So $4^{-3} = \dfrac{1}{4^3}$

   (b) $\dfrac{25}{4}$ **(c)** 4   **(d)** 1

5. **(a)** Equation   **(b)** Identity   **(c)** Identity
   (d) Equation   **(e)** Equation

Module 9: Algebraic Expressions

1. $8x^2 - 20xy + 6xy - 15y^2$ **[1]** $= 8x^2 - 14xy - 15y^2$ **[1]**

2. $p(4p^2 - 1)$ **[1]** $= 4p^3 - p$ **[1]**

   You can multiply the first two terms or the final two terms first; you should still end up with the right answer.

3. $6x - 3 + 4x + 32 + 5$ **[1]** $= 10x + 34$ **[1]** $= 2(5x + 17)$ **[1]**

4. $(a - 16)(a - 3)$ **[2]**

5. $\dfrac{(x-9)(x+2)}{(x+9)(x-9)}$ **[2]** $= \dfrac{x+2}{x+9}$ **[1]**

   In questions like these, expect a factor to cancel. So if the denominator is $(x + 9)(x - 9)$, you can expect a factor in the numerator to be either $x + 9$ or $x - 9$.

6. $\dfrac{(a+3)-(a+2)}{(a+2)(a+3)}$ **[1]**

   Don't forget to put brackets round the second term in the numerator here. An easy thing to forget.

   $= \dfrac{1}{(a+2)(a+3)}$ **[1]**

7. $\dfrac{-x-4}{x+1}$ **[1 for top of fraction and 1 for bottom of fraction]**

8. $(x+2)(x-3)(x+4) = (x+2)(x^2+x-12)$
   $= x(x^2+x-12) + 2(x^2+x-12)$
   $= x^3 + x^2 - 12x + 2x^2 + 2x - 24$
   $= x^3 + 3x^2 - 10x - 24$ **[1 for each correct term]**

9. $\dfrac{x-9}{6} \div \dfrac{x^2-9x}{3} = \dfrac{x-9}{6} \times \dfrac{3}{x^2-9x}$ **[1]** $= \dfrac{x-9}{6} \times \dfrac{3}{x(x-9)}$ **[1]**

   $= \dfrac{1}{2x}$ **[1 for $\dfrac{1}{2}$ and 1 for x in denominator]**

10. $(3x - 5)(x + 8)$ **[2]**

11. $(2x + 1)^3 = (2x + 1)(4x^2 + 4x + 1)$ **[1]**
    $= 8x^3 + 12x^2 + 6x + 1$ **[2] [–1 for an error or omission]**

12. $\dfrac{x^2 - 5x - 84}{x^2 + 5x - 14} = \dfrac{(x+7)(x-12)}{(x+7)(x-2)}$ **[2]** $= \dfrac{x-12}{x-2}$ **[1]**

Module 10: Algebraic Formulae

1. **(a)** 36   **(b)** 48   **(c)** $2\dfrac{2}{3}$
   (d) $-7\dfrac{2}{3}$ **(e)** 6

2. $xy - 8y = 2 + 3x$ **[1]**
   $xy - 3x = 2 + 8y$
   $x(y - 3) = 2 + 8y$ **[1]**
   $x = \dfrac{2 + 8y}{y - 3}$ **[1]**

3. $s - ut = \dfrac{1}{2}at^2$ **[1]**

   $a = \dfrac{2(s - ut)}{t^2}$ **[1]**

4. $y - 5 = \dfrac{2}{x}$ **[1]**

   $x = \dfrac{2}{y - 5}$ **[1]**

5. $q = r - p$ **[2]**

   Try to write the given information mathematically first. This question is basically telling you that $p + q = r$, then asking you to rearrange the equation.

6. **(a)** 31.1 degrees Celsius (1 d.p.)
   (b) $9T_C = 5T_F - 160$ **[1]**

   $T_F = \dfrac{9T_C}{5} + 32$ **[1]**

   (c) 140 degrees Fahrenheit

Module 11: Algebraic Equations

1. $5x - 10 - 3x + 12 = 4$ **[1]**
   $2x = 2$
   $x = 1$ **[1]**

2. $x^2 - 3x - 28 = 0$
   $(x + 4)(x - 7) = 0$ **[1]**
   $x = -4$ **[1]**, $x = 7$ **[1]**

3. $(3x - 5)(x + 2) = 0$ **[1]**
   $x = \dfrac{5}{3}$ **[1]**, $x = -2$ **[1]**

4. $x = 5$ **[1]**, $y = 2$ **[1]**

5. $(x - 4)^2 - 20 = 0$ **[2]**
   $x = 4 \pm \sqrt{20}$ **[1]**

6. $3\left[x^2 + 8x + \dfrac{40}{3}\right]$ **[1]** $= 3\left[(x+4)^2 - \dfrac{8}{3}\right]$ **[1]** $= 3(x+4)^2 - 8$ **[3]**

7. $x = \dfrac{8 \pm \sqrt{220}}{6}$ **[1]**
   $x = -1.14$ **[1]**, $x = 3.81$ **[1]**

8. $x^2 - x - 2 = 2x + 2$ **[1]**
   $x^2 - 3x - 4 = 0$
   $(x + 1)(x - 4) = 0$ **[1]**
   $x = -1,\ y = 0$ **[1]**
   $x = 4,\ y = 10$ **[1]**
   Distance $= \sqrt{5^2 + 10^2} = \sqrt{125}$ **[1]**

9. $8 - x^2 = x - 4$

$x^2 + x - 12 = 0$ **[1]**

$(x+4)(x-3) = 0$ **[1]**

$x = -4$ **[1]**, $x = 3$ **[1]**

$y = -8$ **[1]**, $y = -1$ **[1]**

10. $2x^2 - 11x + \dfrac{9}{2} = 0$ **[2]**

$2x^2 - 10x + 7 = x + \dfrac{5}{2}$ **[2]**

So need to plot $y = x + \dfrac{5}{2}$

> Remember to start with the equation you are trying to **solve**, and add or subtract terms from both sides to obtain the equation you **have**.

11. List of at least five iterations **[2]** leading to $x = 2.49$ **[1]**

Module 12: Algebraic Inequalities

1. $7x > 35$ **[1]** $\quad\quad x > 5$ **[1]**

2. **(a)** $x \leqslant -1$ **(b)** $x > 3$ **(c)** $-4 \leqslant x < 2$

3. $-3, -2, -1, 0, 1, 2, 3$ **[2]**

4. $(x-8)(x-1) \leqslant 0$ **[1]**

[1, 8] **[1 for brackets and 1 for correct values]**

5. **(a)**

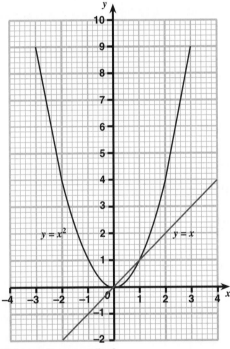

[1 for each correct line]

(b) $0 < x < 1$

(c) $(-\infty, 0) \cup (1, \infty)$ **[2]**

> Even if the sketch was not asked for in this question, it would be useful to draw one.

6. $2 \leqslant x \leqslant 5$ and $1 \leqslant y \leqslant 6$ **[2 for correct values and 2 for correct boundaries]**

7.

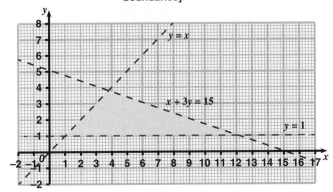

[1 for each correct (dashed) line and 1 for triangle shaded]

> In questions like this, rearrange each equation into the form $y = mx + c$. It will make each line easier to sketch or plot.

Module 13: Sequences

1. 5, −4 **[1 for each correct term]**

2. 44, 57 **[1 for each correct term]**

3. 1, 3, 13, 63, 313 **[2 if fully correct; 1 if one error]**

4. **(a)** $U_n = 4n + 7$ **[1 for 4n; 1 for 7]**

(b) Yes **[1]** since there is a whole number solution to $4n + 7 = 303$

$(n = 74)$ **[1]**

5. **(a)** $a + b + c = 7$

$4a + 2b + c = 6$ **[1]**

$9a + 3b + c = 7$

$5a + b = 1$

$3a + b = -1$ **[1]**

$a = 1$, $b = -4$, $c = 10$ **[1 for each correct value]**

> Always write down the three equations, then subtract the second from the third, and the first from the second.

(b) $U_{20} = 330$

6. **(a)** $U_n = \left(2\sqrt{3}\right)^n$ **[2]** **(b)** $U_3 = 24\sqrt{3}$

Module 14: Coordinates and Linear Functions

1. Line A: 6 **[1]** Line B: −3 **[1]** Line C: $\dfrac{-2}{3}$ **[1]** Line D: 1 **[1]**

2. Lines A and C are parallel. **[1]**

Lines B and D are perpendicular. **[1]**

3. $x = 0$

4. $y = 9 - 6x$ **[1 for 9; 1 for −6x]**

5. $y = 5x + c$ **[1]**

$1 = -15 + c$ **[1]**

$c = 16$

$y = 5x + 16$ **[1]**

6. $m = \dfrac{-5 - (-10)}{6 - (-4)}$ **[1]**

$m = \dfrac{1}{2}$

$y = \dfrac{1}{2}x + c$ **[1]**

Substitute in one coordinate to give $c = -8$ **[1]**

$y = \dfrac{1}{2}x - 8$ **[1]**

> Always show your working out for the gradient. Even if the final answer is wrong, you will still pick up some marks.

7. $y = \dfrac{1}{3}x + 5$ **[1]**

So need $m = -3$

$y = -3x + c$ **[1]**

$-4 = -15 + c$ **[1]**

$c = 11$ **[1]**

$y = 11 - 3x$ **[1]**

Module 15: Quadratic Functions

1. $x = \dfrac{-b}{2a} = \dfrac{-(-30)}{10}$ **[1]**

$x = 3$ **[1]**

2. $(x+4)(x-11) = 0$ **[1]**

$x + 4 = 0$ or $x - 11 = 0$

$x = -4$ **[1]** or $x = 11$ **[1]**

3. $(3x-2)(x+4) = 0$

$3x - 2 = 0$ or $x + 4 = 0$ **[1]**

$x = \dfrac{2}{3}$ **[1]** or $x = -4$ **[1]**

4. (a) $x^2 + 6x + 4 = (x+3)^2 - 5$ **[2]**

So minimum point occurs at (–3, –5) **[1]**

(b) (0, 4)

5. (a) $(x-2)(x-4) - 48 = x^2 - 6x - 40$ **[1]**

$= (x+4)(x-10)$ **[1]**

So crosses axes at $x = -4$ and $x = 10$ **[2]**

(b) (0, –40)

(c) $x = 3$

6. (a) $3x^2 - 12x + 22 = 3\left(x^2 - 4x + \dfrac{22}{3}\right)$ **[1]**

$= 3\left((x-2)^2 - 4 + \dfrac{22}{3}\right)$ **[1]**

$= 3\left((x-2)^2 + \dfrac{10}{3}\right) = 3(x-2)^2 + 10$ **[3]**

(b) (2, 10)

> This is a common type of question. Always factor out the x^2 coefficient completely; complete the square, then multiply through by the number again.

7. $25 + 10x - x^2 = -\left(x^2 - 10x - 25\right)$ **[1]**

$= -\left((x-5)^2 - 50\right)$ **[1]** $= 50 - (x-5)^2$ **[1]**

Maximum point occurs at (5, 50) **[1]**

> Even if the number before the x^2 term is negative, still factor it out completely as in the first line here.

Module 16: Other Functions

1. (a) $y = 5x + 3$

$x = \dfrac{y-3}{5}$ **[1]**

$f^{-1}(x) = \dfrac{x-3}{5}$ **[1]**

(b) $ff(x) = 5(5x+3) + 3$ **[1]** $= 25x + 18$ **[1]**

2.

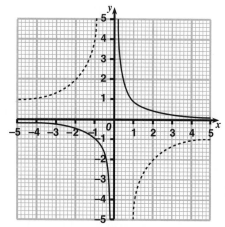

Solid curve $y = \dfrac{1}{x}$ **[1 for correct shape and 1 for correct position]**

Dashed curve $y = \dfrac{-4}{x}$ **[1 for correct shape and 1 for correct position]**

> When sketching reciprocal graphs, make sure the curves never touch or cross the axes.

3. (a) –4   **(b)** –3   **(c)** $\dfrac{-1}{2} - (-2)$ **[1]** $= \dfrac{3}{2}$ **[1]**

4.

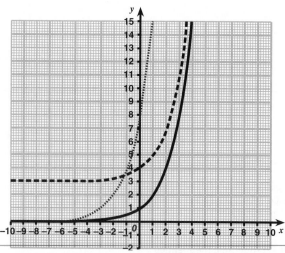

(a) See solid line above. **[1 for graph shape and 1 for line passing through the point (0, 1)]**

(b) See dashed line above. **[1 for graph shape and 1 for line passing through the point (0, 4)]**

(c) See dotted line above. **[1 for graph shape and 1 for line passing through the point (0, 8)]**

> If the curve you are sketching crosses any axes, always make clear where it does so.

5. A translation of 90° in the direction of the positive x-axis (to the right) **[2]** and a translation of 2 units in the direction of the positive y-axis (up) **[2]**.

6. $y = \dfrac{3x+1}{1-x}$

$y - xy = 3x + 1$ **[1]**

$y - 1 = 3x + xy$ **[1]**

$y - 1 = x(3 + y)$ **[1]**

$x = \dfrac{y-1}{3+y}$ **[1]**

$P^{-1}(x) = \dfrac{x-1}{3+x}$ **[1]**

7. $x^2 - 2x + 5 = (x-1)^2 + 4$ **[2]**

Translation of $y = x^2$ **[1]** 1 unit to the right **[1]** followed by translation **[1]** 4 units up **[1]**.

> This could also be described as a single translation through the column vector $\begin{pmatrix} 1 \\ 4 \end{pmatrix}$

Module 17: Problems and Graphs

1. $x^2 + y^2 = \left(\sqrt{2}\right)^2$

so shape is a circle **[1]**, centre (0, 0) **[1]** of radius $\sqrt{2}$ **[1]**

2. (a) $x^2 + y^2 = \dfrac{9}{4}$

$x^2 + y^2 = \left(\dfrac{3}{2}\right)^2$ **[1]**

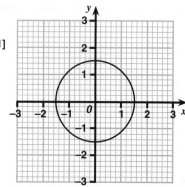

[2]

(b) Equation of tangent is $\frac{1}{2}x + \sqrt{2}y = \frac{9}{4}$ **[2]**

$2x + 4\sqrt{2}y = 9$ **[1]**

$2x + 4\sqrt{2}y - 9 = 0$ **[1]**

3. (a) Area of triangle 1 is $\frac{1}{2} \times \frac{1}{2} \times 5 = \frac{5}{4}$

Area of trapezium 2 is $\frac{1}{2} \times \frac{(5+8)}{2} = \frac{13}{4}$

Area of trapezium 3 is $\frac{1}{2} \times \frac{(8+9)}{2} = \frac{17}{4}$

Area of trapezium 4 is $\frac{1}{2} \times \frac{(8+9)}{2} = \frac{17}{4}$

Area of trapezium 5 is $\frac{1}{2} \times \frac{(5+8)}{2} = \frac{13}{4}$

Area of triangle 6 is $\frac{1}{2} \times \frac{1}{2} \times 5 = \frac{5}{4}$

[4 for calculating areas of the six strips]

The total area is therefore

$\frac{5}{4} + \frac{13}{4} + \frac{17}{4} + \frac{17}{4} + \frac{13}{4} + \frac{5}{4} = \frac{70}{4} = \frac{35}{2}$units² **[2]**

Show clearly how you have calculated each separate area. You are less likely to make a mistake this way.

(b) Draw a suitable tangent line at (0, 0): **[2]**

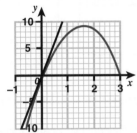

Choose any point lying on the tangent line, e.g. $\left(\frac{1}{2}, 6\right)$ **[1]**

Gradient from (0, 0) to $\left(\frac{1}{2}, 6\right)$ is 12 **[1]**

Equation of tangent is therefore $y = 12x$ **[1]**

4. Substituting (h, k) into the equation $x^2 + y^2 = a^2$ gives $h^2 + k^2 = a^2$ **[1]**

The radius line from (0, 0) to (h, k) has gradient $\frac{k}{h}$ **[1]**

The gradient of the tangent is therefore $\frac{-h}{k}$ **[1]**

The equation of the tangent is therefore $y = \frac{-h}{k}x + c$ **[1]**

Substituting $x = h$ and $y = k$ gives $k = \frac{-h^2}{k} + c$ **[1]**

So $c = \frac{h^2}{k} + k = \frac{h^2 + k^2}{k}$

So $c = \frac{a^2}{k}$

Therefore equation of tangent is $y = \frac{-h}{k}x + \frac{a^2}{k}$ **[1]**

Or $hx + ky = a^2$

Module 18: Converting Measures

1. 0.85×1.25 (or 850×1250) **[1]**

$= 1.06$m² (3 s.f.) (or $1\,062\,500$mm²) **[1]**

2. 3.4×1000^2 **[1]** $= 3\,400\,000$m² **[1]**

3. $100 \times 100 \times 100$cm³ $= 1$m³ **[1]**

$1\,000\,000$g $= 1000$kg **[1]**

4. $68 \div 5 \times 8$ or $110 \div 8 \times 5$ **[1]**

108.8 or 68.75

110km/h is faster **[1]**

5. $24 \times 60 \times 60$m/hour **[1]** $\div 1000$ **[1]** $= 86.4$km/h **[1]**

Multiply m/s by 60×60 to convert to m/hr.

6. (a) Bob $626 \div 35 = £17.89$/h **[1]**

Leila $30\,000 \div (32 \times 52)$ **[1]** $= £18.03$/h

Shahida has best hourly rate. **[1]**

(b) Shahida $18.4 \times 30 \times 52 = £28\,704$

Bob $626 \times 52 = £32\,552$

[1 for either Bob or Shahida correct]

Bob earns most per year. **[1]**

1 year = 52 weeks

7. (a) $463 \div 24$ **[1]** $= 19.29$g/cm³ (2 d.p.) **[1]**

(b) $19.29\ldots \times 30$ **[1]** $= 578.75$g **[1]**

(c) $(463 + 578.75) \times 23.50$ **[1]** $= £24\,481.13$ **[1]**

[No mark for 24 481.125]

Module 19: Scales, Diagrams and Maps

1. (a) $3 \times 20 = 60$m

(b) $57 \div 20$ **[1]** $= 2.85$m **[1]**

Scale 1 : n means 1 and n given in cm. Convert n to m or km as necessary.

2. (a) 5.3cm to 5.5cm

(b) $5.3 \times 5\,000\,000$ **[1]** $= 265$km **[1]** **[Accept up to 275km for a 5.5cm measurement]**

(c) $203° (\pm3°)$ **[2]**

[1 mark for correct construction or 157°]

(d) $(203) - 180$ OR $(203) + 180 - 360$ **[1]** $= 023°$ **[1]**

3. (a)

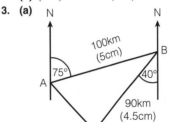

[1 for correct drawing; 1 for accuracy]

(b) $318° (\pm3°)$ **[1]**

58km $(\pm2$km) **[1]**

All bearings are measured clockwise from North line.

4. (a) $18 \times 25\,000$ **[1]** $= 450\,000$cm $= 4.5$km **[1]**

(b) **[2]**

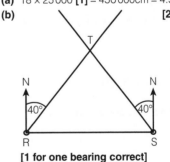

[1 for one bearing correct]

Module 20: Comparing Quantities

1. $\frac{14}{5}$ 37.5% $3\frac{5}{8}$

[2 if fully correct; 1 for two correct]

2. 20%   0.21   0.211 $\frac{2}{9}$ $\frac{3}{10}$

[2 if fully correct; 1 for four in correct order]

Change all amounts to decimals to find the correct order.

3. $2.8 - 2.5 = 0.3$

$2.5 - 2.125 = 0.375$ **[1]**

$0.3 < 0.375$ so $2\frac{4}{5}$ closer **[1] [or equivalent working in fractions]**

4. $\frac{3}{8} \times \frac{5}{6}$ **[1]** $= \frac{5}{16} = 31.25\%$ **[1]**

    Or $\frac{5}{6}$ of 12 = 10 **[1]**

    $\frac{10}{32} = 31.25\%$ **[1]**

5. **(a)** 2.54 ÷ 3.45 OR 5.99 ÷ 3.45 **[1]** → 73.6% (1 d.p.) **[1]**
    (b) 3.45 × 1.15 **[1]** (= £3.97)
    (3.45 × 1.15) × 80 **[1]** (= £317.40)
    55 × 6.99 + 25 × 4 **[1]** = £484.45
    £484.45 − £317.40 = £167.05
    167.05 ÷ 317.40 = 52.6% **[1]**

    Percentage profit = (profit ÷ original amount) × 100

6. **(a)** 0.4 × 0.55 **[1]** + 0.65 × (1 − 0.55) **[1]** = 51.25% **[1]**
    (b) $\frac{35}{100}$ **[1]** $= \frac{7}{20}$ **[1]**

Module 21: Ratio

1. 20cm : 3m **[1]** simplifies to 2 : 30 **[1]** All the rest simplify to 2 : 3

    Divide all parts of the ratio by a common factor.

2. **(a)** 1 + 2 + 7 **[1]** = 10
    2 litres ÷ 10 **[1]** = 200ml
    Navy blue: 200ml; grey: 400ml; white: 1400ml **[1]**
    (b) 1200 ÷ 2 = 600ml and 6000 ÷ 7 = 857ml so grey is
    limiting colour **[1]**
    1 lot is 600ml so total 600 × 10 = 6000ml = 6 litres **[1]**

3. A is $\frac{5}{8}$ of B

4. **(a)** 50 ÷ 5 = 10 so 3 biscuits need 10g.
    3 × 7 = 21 biscuits
    [1 for any correct method; 1 for correct answer]
    (b) 175 ÷ 5 = 35 so 35g flour for 3 biscuits.
    35 × 4 = 140g
    [1 for any correct method; 1 for correct answer]

5. No
    Blue costs 30 ÷ 5 **[1]** = £6/litre and yellow 28 ÷ 7 = £4/litre
    Cost for 7 litres 5 × 4 + 2 × 6 **[1]** = £32
    Lucy gets £4.50 × 7 = £31.50. So she loses 50p **[1]**

Module 22: Proportion

1. 59 × 2.5 **[1]** = 147.5kcal **[1]**
2. **(a)** 500 × 1.29 = €645
    (b) (645 − 570) ÷ 1.33 **[1]** = £56.39 **[1]**

    Each £ receives €1.29 → multiply. Each €1.33 receives
    £1 → divide.

3. **(a)** 1360 ÷ 80 **[1]** = 17 litres/cow
    17 × (80 + 25) **[1]** = 1785
    1785 × 365 × 0.3 **[1]** = £195 457.50 **[1]**
    (b) 800 ÷ 105 = 7.6 days
    [1 for any correct method; 1 for correct answer]

    6 tonnes is 800 cattle feeds.

4. Scale factor × 1.5 **[1]**
    PR = 6 ÷ 1.5 **[1]** = 4cm
    TU = 4.5 × 1.5 = 6.75cm
    [1 for both correct values]

    Scale factor = $\frac{9}{6}$ = 1.5

5. $\frac{270}{120} = \frac{9}{4}$ so scale factor for length = $\frac{3}{2}$ **[1]** and for

    volume = $\frac{27}{8}$

    Volume of P = 2700 ÷ $\frac{27}{8}$ **[1]** = 800cm³ **[1]**

Module 23: Rates of Change

1. **(a)** 10 000 × 1.02⁴ or other stepwise method **[1]**
    = £10824.32 **[1]**
    [1 for 10 000 × 1.02]

    (b) 10 000 × 1.15 **[1]** × 1.06 × 0.82 × 1.01 **[1]**
    = £10 095.76 **[1]**

2. No **[1]**
    6500 × 1.03 = 6695 so £195 is correct for first year but the money
    invested for year 2 is 6500 + 195 so it will earn more interest **[1]**.
    Actual interest = 6500 × 1.03³ − 6500 = £602.73 **[1]**

3. **(a)** 20cm   **(b)** Between 4 and 8 minutes   **(c)** 20 ÷ 4 = 5cm/min

    Steepest gradient gives fastest speed.

4. **(a)** 210 ÷ 10 **[1]** = 21m/s (±0.4m/s) **[1]**
    (b) Drawing tangent at t = 5 **[1]**
    Attempt at gradient **[1]**
    13m/s (±2m/s) **[1]**

Module 24: Constructions

1. **[1 for correct construction lines shown; 1 for accuracy]**

    Always leave your construction lines and arcs, the
    examiner needs to see them.

2. **[1 for width 3.5cm; 1 for accurate lengths and 90° angles]**
3. **[1 for correct construction; 1 (conditional on construction)
    for angle of 90° (±2°)]**
4.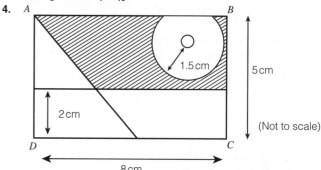

    **[1 for circle radius 1.5cm round tree; 1 for line 2cm from DC; 1
    for correct area shaded]**
5. **[1 for equilateral triangle used to construct 60°; 1 for construction
    of angle bisector; 1 (conditional on construction) for angle of
    30° (±2°)]**

Module 25: Angles

1. For tessellation, the angles must sum to 360°.
    Internal angle of a regular pentagon = 108°
    360 ÷ 108 is not a whole number.
    [1 for 108°; 1 for written or diagrammatic explanation]

2. 180 − 157.5 **[1]** = 22.5
    360 ÷ 22.5 = 16 **[1]**

    It is always the **external** angles that sum to 360°.

3. Internal angle of hexagon = 120°
    Internal angle of octagon = 135°
    [1 for one correct method]
    360 − (120 + 135) **[1]** = 105° **[1]**

4. Triangle ABC isosceles
    Angle ACB = 180 − 2 × 70 **[1]** = 40°
    Angle FCD = 40° (opposite angles)
    Angle FED = 360 − (130 + 130 + 40) = 60° **[1]**
        (angles in a quadrilateral sum to 360°)
    [1 for all four underlined reasons]

    Give a reason for each step of your working.

5. e.g. ∠BPQ = 65° (opposite angles equal)
    ∠PRQ = 180 − 120 = 60° (angles on straight line sum to 180°)
    ∠CPR = 60° **[1]** (alternate angles are equal)
    [1 for any correct, complete reasons]

Module 26: Properties of 2D Shapes

1. A – equilateral, B – obtuse isosceles, C – obtuse scalene,
    D – right-angled isosceles, E – scalene
    [3 if fully correct; 2 if three correct; 1 if one correct]

2. $\angle QBC = c$ (alternate angles are equal) **[1]**
$\angle PBA = a$ (alternate angles are equal) **[1]**
$a + b + c = 180°$ (angle sum on straight line = 180°) **[1]**
Therefore angles in a triangle sum to 180°.

> A proof must work for every possible value so you will need to use algebra.

3. All angles at centre are equal and $360 \div 8 = 45°$
All triangles isosceles so base angle = $(180 - 45) \div 2$ **[1]** = 67.5°
1 internal angle = 2×67.5 **[1]** = 135°
Total = $135 \times 8 = 1080°$ **[1]**
4. Kite
5. Internal angle of an octagon = 135°
Internal angle of an equilateral triangle = 60° **[1 for either 135 or 60]**
Angles at a point = 360° so $x = 360 - (135 + 90 + 60)$ **[1]** = 75° **[1]**

Module 27: Congruence and Similarity
1. B and E
2. G and I

> Congruent = same shape **and** size

3. $\angle PQS = \angle QSR$ (alternate angles) **[1]**
$\angle PSQ = \angle SQR$ (alternate angles) **[1]**
QS common to both triangles, so congruent (ASA) **[1]**
4. Yes, both right angled with sides 3, 4, 5cm **[1]**, so SSS (or RHS) **[1]**
5. Scale factor for length = $\times 1.5$ **[1]** so scale factor for volume = 1.5^3 **[1]**
320×1.5^3 **[1]** = 1080cm³ **[1]**

> (Scale factor for length)² = scale factor for area
> (Scale factor for length)³ = scale factor for volume

Module 28: Transformations
1. **(a)** Reflection **[1]** in $y = x + 1$ **[1]**
(b) Reflection in $x = -1$ **[2 if fully correct; 1 if line drawn only]**
2. Enlargement **[1]**, scale factor $\frac{1}{2}$ **[1]**, centre (0, 0) **[1]**

> Write one fact for each mark.

3.

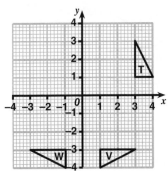

(a) Correct plot
(b) **[2 if fully correct; 1 for rotation]**

> Make sure you do the transformation on the correct shape.

(c) **[2 if fully correct; 1 for reflection of V if it was incorrect in part (b)]**
(d) Reflection **[1]** in line $y = -x$ **[1]**
4.

(a) **[2 if fully correct; 1 for correct scale factor × 2]**

(b) **[2 if fully correct; 1 for correct scale factor × 0.5]**
(c) **[2 if fully correct; 1 for correct scale factor × –1]**

Module 29: Circles
1. **(a)** radius **(b)** diameter
(c) tangent **(d)** circumference
(e) arc **(f)** sector
(g) segment **(h)** isosceles
2. **(a)** False. If PR was a diameter then angle at Q would be 90°.

> If you do not give a correct reason you will not get a mark, even if you have the correct answer.

(b) True. Opposite angles in a cyclic quadrilateral sum to 180°.
(c) False. $\angle SPR = 180 - (50 + 86)$ **[1]** = 44°
If SV is a tangent then angle RSV = angle SPR (alternate segment theorem) **[1]**.
3. **(a)** $\angle PRS = 58°$ **[1]**. Angles at the circumference on the same arc are equal **[1]**.
(b) $\angle QPR = 24°$ (angles at the circumference on the same arc are equal) **[1]**
$\angle PQR = 90°$ (angle on a diameter)
$\angle PRQ = 180 - (90 + 24) = 66°$ (angles in a triangle sum to 180°) **[1]**
$\angle PRQ = 66°$ **[1]**
4. **(a) (i)** $AX = DX$ so triangle DXA isosceles.
$\angle DAX = (180 - 46) \div 2$ **[1]** = 67°
$\angle ABD = \angle DAX = 67°$ (alternate segment theorem) **[1]**
(ii) $\angle ADX = 67°$ and $\angle BDY = 67°$ (alternate angles) **[1]**
$\angle BDA = 180 - (2 \times 67) = 46°$ (angles on a straight line sum to 180°) **[1]**
(b) $\angle BAD = 180 - (67 + 46) = 67°$ **[1]**
Therefore isosceles, as two angles equal **[1]**.
(c) $\angle CFD = \angle BFA$ (opposite angles) **[1]**
$\angle ACD = \angle ABF = 67°$ (angles on same arc are equal)
Angles in both triangles are the same, so they are similar **[1]**.

Module 30: Properties of 3D Shapes
1. Pentagonal-based **[1]** pyramid **[1]**
2. A square-based pyramid **[1]** and triangular prism **[1]** accurately represented by diagrams **[1]**
3.

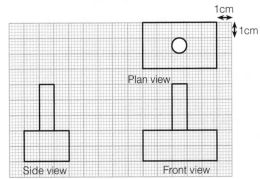

[1 for each drawing; 1 for accuracy]

4. **[3]**

Module 31: Perimeter, Area and Volume
1. **(a)** Area: 8×3 **[1]** = 24m² **[1]**
Perimeter: 26m **[1]**
(b) Area: $\frac{1}{2} \times 14 \times 3$ **[1]** = 21cm² **[1]**
Perimeter: 22.6cm **[1]**
2. $200 \times 80 - 25 \times 20$ **[1]** = 15 500cm² **[1]** (or equivalent calculation to 1.55m²)
3. $\pi \times 9$ **[1]** = 28.274… **[1]** = 28cm **[1]**

> Make sure you know (and use) the correct formula.

4. $\pi \times 4^2$ **[1]** $= 16\pi$ **[1]** cm² **[1]**

> Here the diameter is given but you need the radius to work out area.

5. $\dfrac{\pi \times 5}{2} + 8 + \dfrac{\pi \times 5}{2} + 8$ (or equivalent) **[1]** $= 5\pi + 16 = 31.7$cm **[1]**

6. $(\pi \times 4^2 \times 2)$ **[1]** $+ (\pi \times 8 \times 5)$ **[1]** $= 32\pi + 40\pi = 72\pi$cm² **[1]**

7. $3^3 + \dfrac{1}{3}\pi \times 1.5^2 \times 4$ **[1]** $= 36.4$ **[1]** cm³ **[1]**

8. $\dfrac{4}{3}\pi \times 3.5^3 = a^3$ **[1]**

$a^3 = 179.59$ **[1]**

$a = 5.64$cm **[1]**

9. $\dfrac{2}{3}\pi \times 5^3$ **[1]** $= \dfrac{250}{3}\pi$ **[1]** cm³ **[1]**

Module 32: Pythagoras' Theorem and Trigonometry

1. **(a)** $\cos p = \dfrac{2.8}{3.5}$ **[1]** $= 0.8$ **[1]**

$p = 36.9°$ **[1]**

(b) $\tan 51° = \dfrac{KL}{10.5}$ **[1]**

$KL = 12.966$ **[1]** $= 13.0$m **[1]**

2. $\dfrac{\sin A}{6.3} = \dfrac{\sin 63°}{8.1}$ **[1]**

$\sin A = 0.693...$ so $A = 43.9°$ **[1]**

Angle $ABC = 73.1°$ **[1]**

3. $9^2 = 7^2 + q^2$ **[1]**

> Substitute the values in the Pythagoras equation and then rearrange it.

$q^2 = 9^2 - 7^2$ **[1]** $= 32$

$q = \sqrt{32}$ (or $4\sqrt{2}$) **[1]**

4. **(a)** **(i)** $\dfrac{1}{\sqrt{2}}$ or $\dfrac{\sqrt{2}}{2}$ **(ii)** 1

(b) $a = 2 \times \sin 60° = 2 \times \dfrac{\sqrt{3}}{2}$ **[1]** $= \sqrt{3}$cm **[1]**

$b = 2 \times \cos 60° = 2 \times \dfrac{1}{2}$ **[1]** $= 1$cm **[1]**

5. $\sqrt{14^2 + 6^2 + 3^2}$ (or clear intention to find 3D diagonal) **[1]**

$= 15.5$cm **[1]**

> The longest pencil will fit diagonally across the box.

Module 33: Vectors

1. $\mathbf{a} = 2\mathbf{r}$ $\mathbf{b} = -3\mathbf{s}$

$\mathbf{c} = \mathbf{r} + \mathbf{t}$ $\mathbf{d} = 2\mathbf{t} - \mathbf{s} = \mathbf{r} + \mathbf{t} - 2\mathbf{s}$

[1 for each correct answer]

2. **(a)** $\overrightarrow{BE} = -2\mathbf{b}$

> The opposite sides of a parallelogram are described with equal vectors.

(b) $\overrightarrow{AC} = \mathbf{b} + \mathbf{d}$ **[2]**

(c) $\overrightarrow{CE} = \overrightarrow{CB} + \overrightarrow{BE} = -2\mathbf{b} - \mathbf{d}$ **[2]**

3. $\overrightarrow{OR} = 2\mathbf{q}$ **[1]**

$\overrightarrow{PQ} = \mathbf{q} - \mathbf{p}$ **[1]**

$\overrightarrow{PR} = 2\mathbf{q} - \mathbf{p}$ **[1]**

4. If $\overrightarrow{OP} = 2\mathbf{p}$ and $\overrightarrow{OQ} = 2\mathbf{q}$ then $\overrightarrow{PQ} = 2\mathbf{q} - 2\mathbf{p}$ **[1]**

And $\overrightarrow{OA} = \mathbf{p}$ and $\overrightarrow{OB} = \mathbf{q}$ giving $\overrightarrow{AB} = \mathbf{q} - \mathbf{p}$ **[1]**

$\overrightarrow{PQ} = 2\overrightarrow{AB}$ therefore the lines are parallel **[1]**

> Write each vector in terms of the vectors that you are given and then look for a relationship.

Module 34: Experimental and Theoretical Probability

1. **(a)** $\dfrac{4}{14}$ or $\dfrac{2}{7}$ **(b)** $\dfrac{5}{14}$ **(c)** $\dfrac{9}{14}$

2. $\dfrac{1}{4} \times 120$ **[1]** $= 30$ **[1]**

> Each time the probability she will be correct is a quarter, since there are four different suits.

3. $1 - (0.2 + 0.3 + 0.25 + 0.1)$ **[1]** $= 0.15$ **[1]**

4. **(a)** $\dfrac{12}{25}$ or 0.48

(b) $3 + 4 + 12 = 19$ **[1]**

$\dfrac{19}{25} \times 250$ **[1]** $= 190$ **[1]**

5. $\dfrac{3}{4} \times \dfrac{1}{4}$ **[1]** $= \dfrac{3}{16}$ or 0.1875 **[1]**

> P(not 4) AND P(4) gives the probability of a 4 in exactly two spins.

Module 35: Representing Probability

1. **(a)** ε

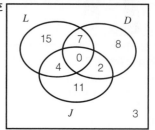

[3 if fully correct; 2 for six correct numbers; 1 for three correct numbers]

> 0 in the middle since no one can vote for all three.
> All the numbers in the diagram must add up to 50.

(b) $\dfrac{11}{50}$ **(c)** $\dfrac{7}{26}$ **[1 for '7' and 1 for '26']**

2. **(a)**

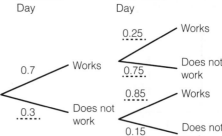

0.3 **[1]**

0.25, 0.75, 0.85, 0.15 **[1]**

(b) P(Christmas AND not New Year) OR P(Not Christmas AND does New Year) =

$(0.7 \times 0.75) + (0.3 \times 0.85)$ **[1]** $= 0.525 + 0.255 = 0.78$ **[1]**

> All four probabilities are different on the second branch because it is conditional on the first branch. Remember to multiply along the branches and then add up the two separate possible outcomes.

Module 36: Data and Averages

1. **(a)** 498   **(b)** 499

> The median is between 498 and 500 and so is 499.

(c) Suitable answers, e.g. The sample size is too small. Need to sample across the range of times during production.

[1 for each reason up to a maximum of 2]

2. **(a)** $60 \leqslant w < 80$

> Median value is between the 80th and 81st value, which is in the $60 \leqslant w < 80$ class interval.

(b) Correct midpoints (35, 45, 55, 70, 90, 110, 135) **[1]**

Correct total of midpoint × frequency of 10 915 **[1]**

Estimated mean = 68.2g **[1]**

(c) $20 + 25 + 22 + 44 = 111$ **[1]**

$111 \div 160 = 69.4\%$ **[1]**

3. 6, 6, 6, 7, 10 or 6, 6, 6, 8, 9 **[3]** Other solutions are possible.
[1 mark for any one of and 2 marks for any two of: a total of 35, median of 6 or mode of 6]

> The total of all five numbers must be 35, i.e. 5 (cards) × 7 (mean) = 35

Module 37: Statistical Diagrams

1. (a) 21
  (b) A bar chart or vertical line graph, with frequency on the y-axis and days of the week on the x-axis, e.g.

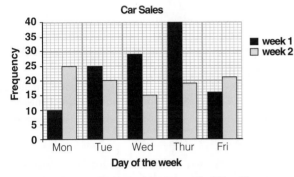

Car Sales

[1 for selecting a dual bar chart or vertical line diagram; 1 for fully correct data (frequencies for each bars); 1 for a fully labelled chart]

2. Table completed as follows: Seafood – 106.7°; Meat – 66.7°; Chicken – 53.3°; Mushroom – 55 and 73.3° **[1 for correct frequency of mushroom and 1 for correct angles]**

> Total frequency must be 45 × 6 = 270 as vegetarian represents one-sixth of the pie chart. 270 – 215 = 55 mushroom

3. (a) Attempt to draw a graph to interpret the data **[1]**
    Correct labelling of axes **[1]**
    Appropriate line of best fit drawn **[1]**
    Estimate 175–195 million **[1]**
  (b) Any suitable answer, e.g. It would not be reliable to use the graph to predict sales in year 14 because it is beyond the range of the data **[1]** and the trend of sales may change **[1]**.

4. (a) A scatter graph drawn with age on x-axis and speed on y-axis **[1]** and the points correctly plotted **[1]**. Data supports John's conclusion. **[1]**
  (b) Negative correlation

> The trend shows that the younger the driver, the higher the speed; and the older the driver, the lower the speed. So the correlation is negative.

5. Attempt to find the frequencies using the histogram **[1]**
    Process to use frequency and midpoints **[1]**
    Method of total (frequency × midpoint) ÷ total frequency **[1]**
    Mean of 36–37 **[1]**
    Correct conclusion that the data supports an increase in travel distances **[1]**

Module 38: Comparing Distributions

1. (a) Class A: 48.8 **[1]**, Class B: 45.7 **[1]**
  (b) Class A: Min 23, LQ 34, Median 45, UQ 62, Max 83 **[2]**
    Class B: Min 27, LQ 32.5, Median 42, UQ 52.5, Max 78 **[2]**

> Order the numbers smallest to biggest. Then find the position of a quarter, half and three-quarters of the way through the data.

  (c) Comparison of either interquartile range (IQR) or median
    Class A: IQR = 62 – 34 = 32; Median 45 **[1]**
    Class B: IQR = 52.5 – 32.5 = 20; Median 42 **[1]**
    Any suitable conclusion, e.g. Class B is more consistent but has a lower median **[1]**.

2. (a) Cumulative frequency curve drawn correctly with points (150, 10); (200, 48); (250, 96); (300, 127); (350, 147); (400, 155); (500, 160) **[3 if fully correct; 2 if one error; 1 for correct cumulative frequencies]**

> Remember cumulative frequency is a running total of the frequency column.

  (b) Minimum = 100; consistent with the curve (values at 40, 80 and 120) for LQ = 190 (±5), median = 230 (±5), UQ = 285 (±5); maximum = 500 **[2 if fully correct; 1 if one error]**

> 160 people in total so LQ will be found at 40th, median at 80th and UQ at 120th. Draw lines across from the y-axis at these points and read off where they hit the curve from the x-axis.

  (c) Any suitable answers, e.g. compare the IQR for both sets of data (approx. 100 and 160) so greater variation for younger drivers; compare the median for both sets of data (230 and 450) so much higher for younger drivers.
    [1 for each comparison]
  (d) Approx. 22 people **[1]**. This is an estimate as the data has been grouped and actual costs are unknown within a group **[1]**.

Exam Practice Paper 1

1. Mean

2. (a) $-5 - 3 = -8$ **[1]**
    $-8 \div 4 = -2$ **[1]**
  (b) $a = -1$ and $b = 7$, $a = -4$ and $b = 4$, $a = -11$ and $b = 2$, $a = -25$ and $b = 1$
    [1 mark for each correct pair of answers]

> Since a must be a negative number, the solution $a = 1$, $b = 14$ is invalid.

3. Ann got two lots more than Timmy.

    1 lot $\dfrac{£32}{2}$ = £16 **[1 for showing correct division]**
    George got (16) × 5 **[1 for your answer × 5]** = £80 **[1]**

4. (a) $\left(\dfrac{20 \times 9}{4}\right) \times 3$ **[1]** = £135 **[1]**

> Round each number to 1 significant figure.

  (b) $885 \times 22 = 19\,470$ **[1]**
    $8.85 \times 22 = £194.70$ **[1]**

5. $8x - 5y = 19$
    $60x + 5y = 15$ **[1]**

    $68x = 34$ **[1]**
    $x = \dfrac{1}{2}$ **[1]**
    $y = -3$ **[1]**

6. P(orange) $= \dfrac{2}{3}$ **[1]**

    $\dfrac{2}{3}$ of 9 = 6 **[1]**

> To start with there was a multiple of 3 chocolates and then a multiple of 4 after 1 had been eaten. So there must have been 9 chocolates to start with.

7. 10% = 34, so 5% = 17 **[1]**
    340 + 34 + 17 = £391 **[1 for correct answer including units]**

8. (a) $\dfrac{4}{6}$ **[1]** $= \dfrac{2}{3}$
  (b) $BC^2 + 4^2 = 6^2$ **[1 for correct use of Pythagoras]**
    $BC = \sqrt{6^2 - 4^2} = \sqrt{20}$

    $\tan B = \dfrac{4}{BC}$ **[1]** $= \dfrac{4}{\sqrt{20}}$ **[1]**

    $= \dfrac{4\sqrt{20}}{\sqrt{20}\sqrt{20}} = \dfrac{\sqrt{20}}{5}$ **[1]** $= \dfrac{2\sqrt{5}}{5}$

9. $\dfrac{15}{16}$

> Turn the second fraction upside down and multiply.

10. $x^2 - 8x - 9 = 11$ **[1]**

$x^2 - 8x - 20 = 0$ **[1]**

$(x + 2)(x - 10) = 0$

$x = -2$ or $x = 10$ **[2]**

11. $81x^8 y^{20}$ **[2]**

12. Using the bar chart to find correct total of years

1, 2, 3, 5 = 95 **[1]**

TV sales: 120 – 95 = 25 (i.e. 25 000) **[1]**

Computer sales: Solving $2a + 5 = 25$, so $a = 10$ (i.e. 10 000) **[1]**

13. $m = 2$, $c = -3$ **[1]**

$y = 2x - 3$ **[1]**

> Gradients of parallel lines are equal.

14. (a) Sides of triangle $x + 3$ **[1]**

Perimeter of triangle = perimeter of square

$3x + 6 = 4x$ **[1]**

$x = 6$ **[1]**

(b) Height of triangle = $\sqrt{9^2 - 3^2}$ **[1 for correct use of Pythagoras on half of triangle]** = $\sqrt{72}$ **[1]**

Diagonal of square = $\sqrt{6^2 + 6^2}$ = $\sqrt{72}$ **[1]** Therefore equal.

15.

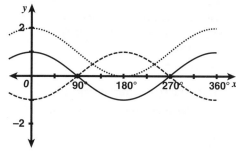

(a) See dashed curve above **[1 for correct shape; 1 for correct position]**

(b) See dotted curve above **[1 for correct shape; 1 for correct position]**

16. $\pi \times 3 \times 2$ **[1]** = 6πcm **[1]**

17. $\angle BAC = 50°$ **[1]** (alternate angles are equal)

$\angle BCA = 180 - (50 + 75) = 55°$ (angle sum of a triangle = 180°)

$\angle BCD = 180 - (55 + 80) = 45°$ (angles on a straight line = 180°)

[1 for correct answer; 1 for three correct reasons]

18. Differences are 5, 11, 17

So next difference is 23 **[1]**

So next term is 54 **[1]**

> You need to go to the second difference to find the pattern.

19. $3x + 5 + x + 8x - 43$ **[1]**

$12x - 38 = 240 - 50$ **[1]**

$x = 19$ **[1]**

$\dfrac{19}{240}$ **[1]**

> Write and solve an equation to find x.

20. After increase, let 1 part be x so values are $5x$ and $2x$

[1 for a method using one ratio]

After decrease, values are $5x – 25$ and $2x – 25$ and these are in ratio 5 : 1

$5x – 25 = 5(2x – 25)$

$100 = 5x$ and $x = 20$ **[1 for correct calculation of a part]**

After increase values are $5 \times 20 = 100$ and $2 \times 20 = 40$

[1 working back to original values]

So original ratio 80 : 20 = 4 : 1 **[1]**

> This could be done by trials, which is quicker!
> 5 : 2...could be £50 : £20 but then original value of B = £0
>   ...could be £100 : £40 and original values £80 : £20 = 4 : 1
>   and after £5 off £75 : £15 = 5 : 1

> An alternative way of solving this question would be:
> $(a + 20) : (b + 20) = 5 : 2$
> Writing the ratio as a fraction gives: $\dfrac{(a + 20)}{(b + 20)} = \dfrac{5}{2}$
> Cross-multiplying gives: $2a - 5b = 60$
> Similarly for $(a - 5) : (b - 5) = 5 : 1$
> Giving equation: $a - 5b = 20$
> These two equations can be solved giving $a = 80$, $b = 20$
> and ratio 4 : 1

21. (a) $4\sqrt{3}$

> Change it to $\sqrt{16} \times \sqrt{3}$

(b) $\dfrac{2}{3 + \sqrt{7}} \times \dfrac{3 - \sqrt{7}}{3 - \sqrt{7}}$ **[1]** = $\dfrac{6 - 2\sqrt{7}}{9 - (\sqrt{7})^2}$ = $\dfrac{6 - 2\sqrt{7}}{2}$ = $3 - \sqrt{7}$ **[1]**

22. 15% of 260 = 26 + 13 = 39 **[1]**

18% of 210 = 21 + 21 – 4.2 = 37.8 **[1 for any correct method]**

So 15% of 260 is greater **[1 for stating the correct answer]**

> In non-calculator percentage questions, there can be several ways to get the correct answer. Here 18% = 20% – 2% or 10% + 5% + 3%. Make sure your working is clear.

23. Any two suitable reasons, e.g. smaller spread so more consistent (interquartile range = 50 vs 155 hours) **[1]**; Higher median (= 370 vs 350) **[1]**

24. Scale factor for volume $\dfrac{810}{240} = \dfrac{27}{8}$ **[1 – you could also write as a ratio 27 : 8]**

Scale factor for length = $\dfrac{\sqrt[3]{27}}{\sqrt[3]{8}} = \dfrac{3}{2}$ **[1 for taking cube roots]**

Scale factor for area = $\dfrac{3^2}{2^2} = \dfrac{9}{4}$ **[1 for squaring for area scale factor]**

Surface area of B = $180 \times \dfrac{9}{4}$ = 405cm² **[1 for correct value]**

25. $x^2 - 8x + 17 = (x - 4)^2 + 1$ **[2]**

Turning point at (4, 1) **[2]**

Crosses y-axis at (0, 17) **[1]**

Line of symmetry $x = 4$ **[1]**

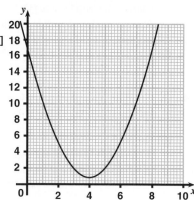

Exam Practice Paper 2

1. $425 \times 0.88 = £374$

[2 for correct answer; 1 for any correct method]

> The multiplier method or working out 12% and subtracting are both correct. If you use a mental method on a calculator paper, you must get the answer correct to gain any marks.

2. 105 **[1]** and 210 **[1]**

$420 = 2 \times 2 \times 3 \times 5 \times 7$ **[1]**

$630 = 2 \times 3 \times 3 \times 5 \times 7$ **[1]**

> Use prime factor trees to find the two common factors
> $2 \times 3 \times 5 \times 7 = 210$ and $3 \times 5 \times 7 = 105$

3. $\pi \times 3.25^2$ **[1]**

$\pi \times 3.25^2 \times \dfrac{3}{4}$ **[1]** = 24.887...

24.9m² **[1 for correct rounding and units]**

4. (a)

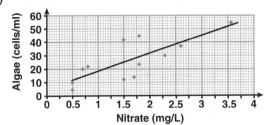

[2 if fully correct; 1 if one error]

(b) Weak positive

(c) Correctly read from line of best fit as shown above [1]: range 1.6–1.9mg/L [1]

(d) Predictions will only be reliable within the range of the nitrate levels in this data (i.e. 0.5–3.6mg/L).

5. (a) $4x^2 - 4x + 1$ [2]

(b) $8x^3 - 12x^2 + 6x - 1$ [2]

6. (a) $\dfrac{41.916\,079\,78}{7.529\,536}$ [1] $= 5.566\,887\,49...$ [1]

> Work out the numerator and denominator separately and write them down.

(b) 5.57 (or your answer to part (a) correctly rounded)

7. $-2 = (-3)^2 + 2(-3) - 5$ [1]
$-2 = 9 - 6 - 5$
$-2 = -2$ [1]

> Substitute $x = -3$ and $y = -2$ into the equation. Remember negative integer rules.

8. $\dfrac{1}{3} \times \dfrac{1}{3}$ [1] $= \dfrac{1}{9}$ [1]

9. $60 : 100 = 3 : 5$ [1 for correct ratio]

$5.4 \times \dfrac{5}{3}$ [1] $= 9$m [1 for answer and units]

> You may use another correct method, e.g. 60cm × 5.4 = 900cm = 9m

10.

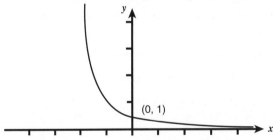

(0, 1)

[2 if fully correct; 1 if curve passes through (0, 1)]

> Always ensure you show clearly where the curve crosses any axes.

11. (a) About 20°C

(b) Any answer in the range 22°C–25°C

(c) $\dfrac{43 - 20}{10} = 2.3$ or $\dfrac{20 - 43}{10} = -2.3$, cooling at rate of 2.3°/minute

[1 for using a tangent on the graph; 1 for correct calculation for your tangent; 1 for answer in range 2–3.3]

12. (a) First farm (no sale): 0.6 [1]
Second farm (from top to bottom): 0.4, 0.6, 0.4, 0.6 [1]

(b) $1 - (0.6 \times 0.6)$ [1] $= 1 - 0.36 = 0.64$ [1]

> Calculate 1 – P(no sale, no sale)

13. $5y - xy = 10x - 2$ [1]
$5y + 2 = 10x + xy$ [1]
$5y + 2 = x(10 + y)$ [1]

$x = \dfrac{5y + 2}{10 + y}$ [1]

14. $\sin R = \dfrac{3.7}{5.8}$ [1] $= 0.6379...$

$\sin^{-1} \dfrac{3.7}{5.8} = 39.6377...$ [1] $= 39.6°$ [1 for correct rounding and degree symbol]

15. (a) $\dfrac{27}{9}$ [1] $= 3$

$9 + 7 = 16$ [1 for adding parts]
$3 \times 16 = 48$kg [1]

(b) 10^6cm³ in 1m³ [1]
8900×1000g [1]
$8900 \times 1000 \div 1\,000\,000$ [1] $= 8.9$g/cm³ [1]

(c) Density $= \dfrac{\text{mass}}{\text{volume}}$ so 9g of copper has a volume of

$\dfrac{9}{8.9}$ [1] $= 1.01$cm³ and 7g of zinc has a volume of 1cm³.

Total mass = 16g
Total volume = 1 + 1.01 = 2.01 [1 for mass and volume]
Density of alloy = 16 ÷ 2.01 = 7.96g/cm³ (2 d.p.) [1]

16. $x^2 - 7x + 14 = 2x - 4$ [1]

$x^2 - 9x + 18 = 0$

$(x - 6)(x - 3) = 0$ [1]

$x = 6$ or $x = 3$

At $x = 6$, $y = 8$ [1] and at $x = 3$, $y = 2$ [1]
So coordinates are (6, 8) and (3, 2)

17. $x = 0.08\dot{1}$
$10x = 0.8\dot{1}$
$1000x = 81.8\dot{1}$
$990x = 81$ [1]

$x = \dfrac{81}{990}$ [1] $= \dfrac{9}{110}$ [1]

> Label the recurring decimal as x and then multiply by 10 and 1000, so that when you subtract them the answer will be an integer, as the recurring decimals have cancelled each other out.

18. New gradient is $\dfrac{-1}{(-2)} = \dfrac{1}{2}$ [1]

$y = \dfrac{1}{2}x + c$ [1]

Substituting $x = 8$, $y = -1$ [1]
$-1 = 4 + c$
$c = -5$ [1]

$y = \dfrac{1}{2}x - 5$

19. (a) $1 : 9$

(b) $10\,800 \div 27$ [1] $= 400$ml (or 400cm³) [1]

20. $\dfrac{KM}{\sin 100°} = \dfrac{KL}{\sin 35°}$ [1 for correct application of sine rule]

$KL = \dfrac{10}{\sin 100°} \times \sin 35°$ [1] $= 5.82$cm [1]

21. $V_n = 4(5(n + 2) + 2) - 2(5n + 2)$ [1]

$V_n = 4(5n + 12) - 10n - 4$

$V_n = 10n + 44$ [2]

$V_n = an + b$ so is arithmetic sequence [1]

$U_n = V_n \Rightarrow 10n + 44 = 5n + 2$ [1]

No, since no positive integer solution for n [1]

Exam Practice Paper 3

1. (a) Lysette, as she has done the most coin flips.

(b) $\dfrac{28}{80} \times 500$ [1] $= 175$ [1] or $\dfrac{44}{120} \times 500$ [1] $= 183$ [1]

2. $3x + 4 > 19 \Rightarrow x > 5$ [1]

$2x - 1 < 13 \Rightarrow x < 7$ [1]

So $x = 6$ [1]

3. (a)

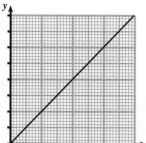

[1 for straight line with positive gradient; 1 for line from (0, 0)]

(b)

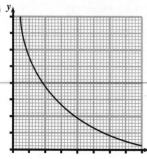

[1 for concave curve; 1 for lines approaching but not touching axes]

4. (a) [1 for correct use of scale]

Answer in range 420–450km [1]

(b) Using your answer from part (a) (for example we are using 440km) $\dfrac{700}{440} = 1.590 = 159\%$ Answer in range 155–165%

[1 for correct division; 1 for correct percentage]

(c) $\dfrac{700}{11} = 63.64 = 64$kmh^{-1} or 64km/h

[1 for method; 1 for correct answer; 1 for correct unit]

5. (a) 3^{-5} **(b)** 7.02×10^{-5} **(c)** 4

6. (a) $40 \leqslant h < 60$

(b) Midpoints (10, 30, 50, 70, 100) [1]

Sum of frequency × Midpoints = 7000 [1]

$7000 \div 140 = 50$cm [1]

(c) Correct calculation of $42 \div 140 = 30\%$ or $\dfrac{3}{10}$ [1].

So the data does not support the farmer's claim. [1]

7. (a) 44 800km/h

(b) $(44\,800 \times 3) \div 2$ [1] $= 67\,200$ [1] $= 6.72 \times 10^4$km [1]

Remember to put your answer back into standard form, with any relevant units.

8. (a) [1 for constructing radii to B and D]

Angle $BOD = 360 - (90 + 90 + 52)$ [1] $= 128°$

Angle $BCD = \frac{1}{2}$ angle $BOD = 64°$ [1]

(b) $CBED$ is cyclic quadrilateral; angle $BED = 180 -$ angle BCD [1]

$= 116°$ [1]

9. (a) £18 000

(b) $18\,000 \times 0.85 \times 0.85$ [1] $= £13\,005$ [1]

(c) $18\,000 \times 0.85^4 = £9396.11$ [1]

4 years [1]

10. $6(2x+2) - 4x = 2x(2x+2)$ [1]

$8x + 12 = 4x^2 + 4x$ [1]

$x^2 - x - 3 = 0$ [1]

In a quadratic such as this, remember you can and should divide through by a common factor to make the numbers easier to deal with.

$x = \dfrac{1 \pm \sqrt{13}}{2}$

$x = 2.30$ [1] or $x = -1.30$ [1]

11.

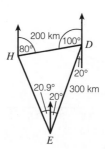

$180 - 100 - 20 = 60°$ [1 for correct method to calculate angle at D]

$EH^2 = 200^2 + 300^2 - 2 \times 200 \times 300 \times \cos 60°$ [1 for correct substitution into cosine rule]

$EH^2 = 70\,000$ so $EH = 264.575...$km [1]

$\dfrac{\sin E}{200} = \dfrac{\sin 60°}{EH}$

$\sin E = \dfrac{\sin 60°}{EH} \times 200$ [1 for correct substitution into sine rule]

$\sin E = 0.6546...$ $E = 40.9°$ [1]

Bearing $= 360 - (40.9 - 20) = 339.1°$ or $339°$ [1]

12. Ratio of radii 4 : 5 so whole cone has height 25cm

[1 for attempt to find height using ratio or scale factors; 1 for finding height]

Volume of frustum $= \frac{1}{3}\pi \times 15^2 \times 25 - \frac{1}{3}\pi \times 12^2 \times 20$ [1]

$= 2874.55$ [1] $= 2870$cm^3 [1 for correct rounding and units]

13. $4x^2 - 24x + 41 = 4\left[x^2 - 6x + \dfrac{41}{4}\right]$ [1]

$= 4\left[(x-3)^2 - 9 + \dfrac{41}{4}\right]$ [1]

$= 4\left[(x-3)^2 + \dfrac{5}{4}\right]$ [1]

$= 4(x-3)^2 + 5$ [1]

$(a = 4, b = 3, c = 5)$

14. (0.7×0.7) [1] $+ (0.3 \times 0.8)$ [1] $= 0.49 + 0.24 = 0.73$ [1]

Draw a tree diagram to help visualise the probabilities.

15. Lower bound for bottle = 747.5ml. Upper bound for cup = 5.5ml [1]

$\dfrac{747.5}{5.5}$ [1] $= 135.909... = 135$ full cups [1]

16. (a) $g(3) = 7$ [1] $f(7) = 20$ [1]

In questions like these, it is often easier to substitute numbers in first, rather than work out an algebraic expression for $fg(x)$

(b) $f^{-1}(x) = \dfrac{1+x}{3}$ [1]

$g^{-1}(x) = \dfrac{x-1}{2}$ [1]

$\left(\dfrac{1+x}{3}\right)\left(\dfrac{x-1}{2}\right) = 4$

$x^2 - 1 = 24$ [1]

$x = \pm 5$ [2]

17. $(16^{\frac{1}{4}})^3 = 2^3$ [1] $= 8$ [1]

18.

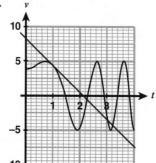

Tangent line drawn [2]

Gradient approximately $\dfrac{-8.7}{2.1}$ [2]

Acceleration is approx. -4.1m/s^2 [1]

When estimating the gradient here, it is easier (and more accurate) if you use the points where the tangent line crosses the axes.